T0406577

Developments in the Flow of Complex Fluids in Tubes

Dennis A. Siginer

Developments in the Flow of Complex Fluids in Tubes

Dennis A. Siginer
Botswana International University
of Science and Technology
Palapye, Botswana

Universidad de Santiago de Chile
Santiago, Chile

ISBN 978-3-319-02425-7 ISBN 978-3-319-02426-4 (eBook)
DOI 10.1007/978-3-319-02426-4
Springer Cham Heidelberg New York Dordrecht London

Library of Congress Control Number: 2014946820

Printed on acid-free paper

Springer is part of Springer Science+Business Media (www.springer.com)

Preface

This monograph together with its complimentary volume [*Siginer, D. A., Stability of Non-linear Constitutive Formulations for Viscoelastic Fluids, Springer, New York,* 2014] in this series is an attempt to give an overall comprehensive view of a complex field, only 60 or so years old, still far from being settled on firm grounds, that of the dynamics of viscoelastic fluid flow and suspension flow in tubes. The monograph on "Stability of Non-linear Constitutive Formulations for Viscoelastic Fluid Media" covers the development of constitutive equation formulations for viscoelastic fluids in their historical context together with the latest progress made, and this volume covers the state-of-the-art knowledge in predicting the flow of viscoelastic fluids and suspensions in tubes highlighting the historical as well as the most recent findings. Most if not all viscoelastic fluids in industrial manufacturing processes flow in laminar regime through tubes, which are not necessarily circular, at one time or another during the processing of the material. Laminar regime is by far the predominant flow mode for viscoelastic fluids encountered in manufacturing processes, and it is extensively covered in this monograph. Turbulent flow of dilute viscoelastic solutions is a topic which has not received much attention except when related to drag reduction. For particle-laden flows there are very interesting developments in both laminar and turbulent regime, and they are duly covered. It is critically important that the flow of non-linear viscoelastic fluids and suspensions in tubes can be predicted on a sound basis, thus the *raison d'être* of this volume. As flow behavior predictions are directly related to the constitutive formulations used, this volume relies heavily on the volume on [*Siginer, D. A., Stability of Non-linear Constitutive Formulations for Viscoelastic Fluids, Springer, New York,* 2014].

The science of rheology defined as the study of ***the deformation and flow of matter*** was virtually single-handedly founded and the name invented by Professor Bingham of Lafayette College in the late 1920s. Rheology is a wide encompassing science which covers the study of the deformation and flow of diverse materials such as polymers, suspensions, asphalt, lubricants, paints, plastics, rubber, and biofluids, all of which display non-Newtonian behavior when subjected to external stimuli and as a result deform and flow in a manner not predictable by Newtonian mechanics.

The development of rheology, which had gotten to a slow start, took a boost during WWII as materials used in various applications, in flame throwers, for instance, were found to be viscoelastic. As Truesdell and Noll famously wrote [*Truesdell, C. and Noll, W., the Non-Linear Field Theories of Mechanics,* 2nd *ed., Springer, Berlin,* 1992] "By 1949 all work on the foundations of Rheology done before 1945 had been rendered obsolete." In the years following WWII, the emergence and rapid growth of the synthetic fiber and polymer processing industries, appearance of liquid detergents, multigrade oils, non-drip paints, and contact adhesives, and developments in pharmaceutical and food industries and biotechnology spurred the development of rheology. All these examples clearly illustrate the relevance of rheological studies to life and industry. The reliance of all these fields on rheological studies is at the very basis of many if not all of the amazing developments and success stories ending up with many of the products used by the public at large in everyday life.

Non-Newtonian fluid mechanics, which is an integral part of rheology, really made big strides only after WWII and has been developing at a rapid rate ever since. The development of reliable constitutive formulations to predict the behavior of flowing substances with non-linear stress–strain relationships is quite a difficult proposition by comparison with Newtonian fluid mechanics with linear stress–strain relationship. The latter does enjoy a head start of two centuries tracing back its inception to Newton and luminaries like Euler and Bernoulli. With the former the non-linear structure does not allow the merging of the constitutive equations for the stress components with the linear momentum equation as it is the case with Newtonian fluids ending up with the Navier–Stokes equations. Thus, the practitioner ends up with six additional scalar equations to be solved in three dimensions for the six independent components of the symmetric stress tensor. The difficulties in solving in tandem this set of non-linear field equations, which may involve both inertial and constitutive non-linearities, cannot be underestimated. Perhaps equally importantly at this point in time in the unfolding development of the science, we are not fortunate enough to have developed a single constitutive formulation for viscoelastic fluids, which may lend itself to most applications and yield reasonably accurate predictions together with the field balance equations. The field is littered with a plethora of equations, some of which may yield reasonable predictions in some flows and utterly unacceptable predictions in others. Thus, we end up with classes of equations for viscoelastic fluids that would apply to classes of flows and fluids, an ad hoc concept at best that hopefully will give way one day to a universal equation, which may apply to all fluids in all motions. In addition the stability of these equations is a very important issue. Any given constitutive equation should be stable in the Hadamard and dissipative sense and should not violate the basic principles of thermodynamics.

Dynamics of tube flow of non-Brownian suspensions and its underpinning field turbulent motion of linear (Newtonian) fluids shows interesting similarities with the flow of viscoelastic fluids in that the secondary flows of viscoelastic fluids in laminar flow driven by unbalanced normal stresses have a counterpart in the turbulent motion of linear fluids in straight tubes of non-circular cross section and in the laminar motion of particle-laden linear fluids. The latter secondary flows are driven by normal stresses due to shear-induced migration of particles. This is a new topic of hot research thrust given its implications in applications. The direction of these normal stresses is opposite of those present in the flow field of a viscoelastic fluid. In fact the tying thread among these seemingly different motions is that all are driven by normal stresses. The turbulent flow of linear fluids is known to have a transversal field due to the anisotropy of the Reynolds stress tensor in non-circular cross sections which entails unbalanced normal Reynolds stresses in the cross section perpendicular to the axial direction. Secondary field also exists in the turbulent flow of linear fluids in circular cross sections if the symmetry is somehow broken due, for example, to unevenly distributed roughness on the boundary, which would again trigger anisotropy of the Reynolds stress tensor. It is not possible to develop a good understanding of the mechanics of the secondary field both in laminar and turbulent motion of particle-laden fluids without a clear grasp of the underlying mechanics of the turbulent secondary field of homogeneous linear fluids. Thus a complete review of both is presented including interesting constitutive similarities with viscoelastic fluids which do arise when certain non-linear closure approximations are made for the anisotropic part of the Reynolds stress tensor.

The impact of the secondary flows on engineering calculations is particularly important as turbulent flows in ducts of non-circular cross section are often encountered in engineering practice. Some examples are flows in heat exchangers, ventilation and air-conditioning systems, nuclear reactors, impellers, blade passages, aircraft intakes, and turbomachinery. If neglected significant errors may be introduced in the design as secondary flows lead to additional friction losses and can shift the location of the maximum momentum transport from the duct centerline. The secondary velocity depends on cross-sectional coordinates alone and therefore is independent of end effects. It is only of the order of 1–3 % of the streamwise bulk velocity, but by transporting high-momentum fluid toward the corners, it distorts substantially the cross-sectional equal axial velocity lines; specifically it causes a bulging of the velocity contours toward the corners with important consequences such as considerable friction losses. The need for turbulence models that can reliably predict the secondary flows that may occur in engineering applications is of paramount importance.

Efforts have not been spared to be thorough in the presentation with commentaries about the successes and failures of each theory and the reasons behind them. The link between different theories and the naturally unfolding succession of theories over time borne out of the necessity of better predictions as well as the challenges in the field at this time are given much emphasis at the expense of a detailed in-depth development of various theories. This book provides a snapshot of a fast developing topic and a bridge connecting new research results with a timely and comprehensive literature review. For a detailed in-depth exposition of anyone subject included in this book, the reader is referred to the extensive reference list.

The responsibility for any mistakes and misquotes that may have crept up into the text in spite of extensive checking remains solely with the author.

As a final note we remark that the intention of this book is to emphasize the common features of and the links between three seemingly different fields, tube flow of viscoelastic fluids, turbulent tube flow of Newtonian fluids, and tube flow of non-colloidal suspensions, and to help bridge the mechanics of all three. The notation used in the literature by the practitioners of all three fields is somewhat different and is kept the same in this book not to create confusion with the existing literature. For instance, the extra stress in viscoelastic fluid flow is denoted by $\mathbf{S}$, whereas the Reynolds stress in the turbulent flow of Newtonian fluids, which is essentially an extra stress as well, is denoted by $\boldsymbol{\tau}$. In the same vein the suspension stress is indicated by $\boldsymbol{\Sigma}$. The symbols are carefully defined wherever they appear in the text.

Palapye, Botswana and Santiago, Chile Dennis A. Siginer

Contents

Chapter 1
Preamble

The focus of this monograph is on the tube and channel flow of non-linear viscoelastic fluids and non-Brownian suspensions in straight tubes of arbitrary but longitudinally constant cross section. The state of the science in predicting the longitudinal field of viscoelastic fluids closely related to the secondary field and the prediction of the friction factors is thoroughly covered. Laminar and turbulent longitudinal flow fields and related phenomena including the all-important drag reduction, the early developments in the history of transversal flows, similarities with secondary transversal deformations associated with the simple shearing of solid materials in non-linear solid mechanics, as well as the analogy between the laminar flow of non-linear fluids and the turbulent flow of linear fluids in non-circular cross-sectional tubes together with constitutive criteria for the existence of secondary flows are commented on as well as the relatively recent research in secondary flows of dilute solutions in rotating pipes and channels and the related drag reduction together with the fundamental aspects of transversal flows and industrial applications. The importance and implications of secondary flows of non-linear viscoelastic fluids and the analytical methods used to investigate secondary flows are discussed. Archival literature concerning the study of secondary flows of non-Newtonian fluids is rather thin when compared to the flurry of research output in other aspects of the behavior of constitutively non-linear liquids in spite of the occurrence and the importance of secondary flows in almost every industrial operation of significance involving constitutively non-linear fluids such as extrusion processes, for instance. A comprehensive and detailed survey of secondary flows of incompressible viscoelastic liquids in straight stationary and rotating tubes is given. Only hydrodynamically and thermally developed flows are considered. Developing flows and the more complex phenomena of secondary flows of complex fluids in curved circular as well as non-circular curved tubes are not included and will be discussed elsewhere.

The stability and drag-reducing characteristics of secondary flows of dilute, weakly viscoelastic solutions in laminar flow in rotating tubes and channels are reviewed as well as the conclusions reached both through analytical and numerical

D.A. Siginer, *Developments in the Flow of Complex Fluids in Tubes*,
DOI 10.1007/978-3-319-02426-4_1

approaches concerning the physics of these flows in straight tubes and its implications for industrial applications. Recent work which explores for the first time the structure of the secondary flow field in quasi-steady flows of constitutively non-linear simple fluids of "multiple integral" type as well as the secondary field in the steady flow of a popular class of non-affine constitutive equations in straight tubes of arbitrary cross sections is summarized.

Tube flow of concentrated suspensions, those with a volume fraction ϕ of more than 20 %, in the laminar regime in non-circular cross sections and in turbulent regime in circular cross sections, is a fascinating subject and shows similarities with the flow of viscoelastic fluids in non-circular tubes in normal stress-driven component of the motion. The mechanics of dilute (volume fraction $\phi < 10$ %) and semidilute (volume fraction 10 % $< \phi < 20$ %) suspensions are reasonably well understood, primarily due to the work of Einstein and Batchelor. However, constitutive equations relating stress to rate of strain for concentrated suspensions $\phi > 20$ % are not generally known, and hence their rheology is still a subject of much investigation despite an emphasis on this issue over the past decades. Microstructure, which refers to the relative position and orientation of physical entities in the material, is the key to understanding the fluid mechanics and rheology of concentrated suspensions. Microstructure is vital to the development of constitutive equations for concentrated suspensions and to the understanding of the viscosity behavior as well as normal stress differences. Recent research demonstrating that secondary flows in particle-laden laminar flow of linear fluids ($\phi > 20$ %) prompted and sustained by shear-driven migration of particles are caused by unbalanced normal stresses, and in particular by particle contributed normal stresses, very much like in the case of polymeric fluids as well as secondary flows of the second kind in turbulent flow of linear fluids in circular cross-sectional straight tubes triggered by a non-uniform distribution of suspended particles or small enough suspended droplets or by non-uniform boundary conditions such as non-uniformly distributed boundary roughness, which all initiate anisotropy in the Reynolds stress tensor, is discussed in detail.

Chapter 2
Longitudinal Flow Field

Abstract The state of the science in predicting the hydrodynamically developed longitudinal field of non-linear viscoelastic fluids in straight tubes of arbitrary but longitudinally constant cross section and in predicting the friction factors is summarized. Laminar and turbulent longitudinal flow fields and related phenomena including recent progress concerning the all-important drag reduction phenomena are addressed.

Keywords Non-linear viscoelastic fluids • Longitudinal velocity field • Straight tubes • Arbitrary cross-section • Friction factors • Laminar flow • Turbulent flow • Pseudoplastic behavior • Drag reduction • Drag-reducing additives and applications • Theoretical mechanism of drag reduction • Experimental findings in drag reduction • Factors influencing the effectiveness of drag reduction

Only hydrodynamically developed flows will be reviewed. It should be noted that hydrodynamically developing flows are important as the entrance length L given in terms of either dimensionless quantities L^+ or L^*,

$$L = L^+ D_h Re^+, \qquad L = L^* D_h Re^*,$$

can be substantial in particular in laminar flow. Here D_h is the hydraulic diameter, and Re^+ and Re^* are the generalized Reynolds number and the Kozicki generalized Reynolds number, respectively, defined in Sect. 2.1.

2.1 Laminar Flow

The comprehensive monograph by Shah and London [1] which covers the research up to 1978 with an extensive bibliography and the review by Shah and Bhatti [2] which appeared a decade later are excellent references for the calculation of the friction factors f for Newtonian fluids in fully established laminar flow in ducts of non-circular shape. The investigations by Eckert and Irvine [3] and Carlson and Irvine [4] on the pressure drop in triangular-shaped ducts are among the earliest on the friction factor in fully established flow of Newtonian fluids in tubes of non-circular

D.A. Siginer, *Developments in the Flow of Complex Fluids in Tubes*,
DOI 10.1007/978-3-319-02426-4_2

cross sections. For the much emphasized rectangular cross sections, Shah and London [1] provide an approximate formula

$$fRe = 24\left(1 - 1.355\alpha^{*} + 1.9467\alpha^{*2} - 1.7012\alpha^{*3} + 0.9564\alpha^{*4} - 0.2537\alpha^{*5}\right) \tag{2.1}$$

where the Reynolds number Re is based on the hydraulic diameter D_{h} and α^* represents the duct aspect ratio with height $2b$ over width $2a$. The approximate formula (2.1) stays within 0.05 % of the correct analytical solution. A simpler but a slightly more approximate formula valid when $\alpha^* \geq 0.05$ whose predictions stay within 0.25 % of those based on the exact analytical velocity w profile,

$$w = \frac{16a^2}{\pi^3\eta}\left(-\frac{\mathrm{d}p}{\mathrm{d}z}\right)\sum_{k=1}^{\infty}\left(\left[\frac{(-1)^{k-1}}{(2k-1)^3}\right]\left\{1 - \frac{\cosh\left[(2k-1)\frac{\pi y}{2a}\right]}{\cosh\left[(2k-1)\frac{\pi b}{2a}\right]}\right\}\cos\frac{(2k-1)\pi x}{2a}\right),$$

was obtained by Natarajan and Lakshmanan [5],

$$fRe = 14.4(\alpha^*)^{-1/6}.$$

In these formulas the pressure, the viscosity, and the longitudinal coordinate are denoted by p, η, and z, respectively. The friction factors for the fully developed laminar flow of power-law fluids in rectangular channels were calculated by Schecter [6] and Wheeler and Wissler [7] via a variational method and a numerical solution, respectively. The latter authors derive an approximate equation for the square duct geometry,

$$4fRe^{+} = 7.4942\left\{\frac{1.7330}{n} + 5.8606\right\}^{n}$$

$$Re^{+} = \frac{\rho\overline{w}^{2n}D_h^n}{K}.$$

where $\overline{w}$ is the average velocity. In fully established laminar flow in circular tubes, the pressure drop of power-law fluids was predicted based on the constitutive assumption, Metzner and Reed [8],

$$\tau = K\left(2tr\mathbf{D}^2\right)^{\frac{n}{2}} \tag{2.2}$$

where τ and $\mathbf{D}$ represent the shear stress and the rate of deformation tensor with K and n referring to the consistency and the power index, respectively. The solution of the momentum equation for the longitudinal flow velocity w with the constitutive assumption Eq. (2.2) gives for the fully established velocity profile w,

$$\frac{w}{w_{\text{max}}} = \frac{3n+1}{n+1}\left\{1 - \left[\frac{r}{R}\right]^{\frac{n+1}{n}}\right\}$$

with w_{max} and R standing for the maximum velocity and the circular tube radius. The friction factor f is defined in terms of a generalized Reynolds number Re',

$$f = \frac{16}{Re'}$$

$$Re' = \rho(w_{\text{max}})^{2-n}(2R)^n\left\{8^{n-1}K\left[\frac{3n+1}{4n}\right]^n\right\}^{-1} \tag{2.3}$$

That this prediction for purely viscous (generalized Newtonian) fluids also holds for viscoelastic fluids in circular pipes was shown by Metzner [9] and Cho and Hartnett [10]. The flow of a generalized Newtonian fluid in ducts of arbitrary cross section was investigated by Kozicki et al. [11] via a generalized Rabinowitsch–Mooney approach. They show in particular that the friction factor for the fully established laminar flow of a power-law fluid in rectangular geometries is given with good accuracy by

$$f = \frac{16}{Re^*},$$

$$Re^* = \rho(w_{\text{max}})^{2-n}D_h^n\left\{8^{n-1}K\left[F(b) + \frac{F(a)}{n}\right]^n\right\}^{-1} \tag{2.4}$$

where $F(b)$ and $F(a)$ are functions of the geometry of the tube, that is, of its aspect ratio α^*, and are tabulated. For instance, for an aspect ratio of $\alpha^* = 0.5$, the values assumed by $F(a)$ and $F(b)$ are 0.2439 and 0.7278, respectively. If $F(a)$ and $F(b)$ are assigned the values of 0.25 and 0.75, respectively, circular duct is obtained and Eq. (2.4) collapses onto Eq. (2.3).

That these predictions based on a purely viscous generalized Newtonian model also hold well for viscoelastic fluids with good accuracy, in particular in rectangular geometries, was proven by the experiments of Hartnett et al. [12], Hartnett and Kostic [13], and Bamidipaty [14] who investigated the flow in a square tube, in a tube of aspect ratio 2 (height is twice as large as the width), and in a tube of aspect ratio 5, respectively, to determine that the overall agreement with power-law predictions is within 10 %. It should be emphasized that given the unavoidable experimental scatter this is a very good agreement. There are no experimental pressure drop measurements available for the fully established laminar flow of viscoelastic fluids in arbitrary cross sections. But on the basis of the evidence available for rectangular cross sections, it makes sense to extend the findings for rectangular cross sections to arbitrary cross sections and assume that unless future experimental evidence proves the contrary pressure drop in the fully established laminar flow of viscoelastic fluids in tubes of arbitrary but constant cross section in the longitudinal direction is well predicted by the generalized Newtonian model.

However, recent research by Siginer and Letelier [15] provides evidence that elastic effects clearly contribute to the longitudinal velocity field, and thus predicting the longitudinal velocity field with power-law type of constitutive equations that does not account for elastic effects, reasonably good that the predictions may be, would give only approximately correct results. Siginer and Letelier [15] have studied the fully developed steady pressure-gradient-driven laminar flow of a class of non-linear and non-affine, single mode, quasilinear viscoelastic fluids with instantaneous elasticity in straight tubes of arbitrary cross section. The class of viscoelastic fluids investigated includes the non-affine Johnson–Segalman [16] and Phan-Thien [18] models and the affine Phan-Thien–Tanner [17] model as special cases (see Sect. 3.6.1 for details). A continuous one-to-one mapping is used to obtain arbitrary tube contours from a base tube contour ∂D_0, Siginer and Letelier [15]. The analytical method presented is capable of predicting the velocity field in tubes with arbitrary cross section. The base flow is the Newtonian field and is obtained at $O(1)$. Field variables are expanded in asymptotic series in terms of the Weissenberg number We. The analysis does not place any restrictions on the smallness of the driving pressure gradients which can be large and applies to dilute and weakly elastic non-linear viscoelastic fluids. The Newtonian field in arbitrary contours is obtained, and longitudinal velocity field components due to shear thinning and to non-linear viscoelastic effects are identified. Third-order analysis shows a further contribution to the longitudinal field driven by first normal stress differences. Their analysis provides evidence that predicting the longitudinal velocity field of non-linear viscoelastic fluids in straight tubes with power-law type of constitutive equations that does not account for elastic effects is only approximate, reasonably good that it may be, for elastic effects clearly contribute to the longitudinal velocity field. Their research, on secondary flows in arbitrary cross-sectional tubes driven by unbalanced second normal stresses in the cross section, is summarized in Sect. 3.6.1. Longitudinal equal velocity contours, the first and the second normal stress differences, as well as wall shear stress variations are discussed for several non-circular contours some for the first time. Examples are shown in Figs. 2.1 and 2.2.

A numerical method to predict the longitudinal velocity field based on the mapping of the simply connected physical flow domain of a straight tube of arbitrary cross section onto a circular computational domain and the efficient and robust Levenberg–Marquardt algorithm, Chai and Yeow [19], has been devised by Ahmeda et al. [20]. The approach they develop permits simple discretization schemes in the circular computational domain leading to simultaneous determination of the velocity and stress contours. The method is applied to the flow of Newtonian, purely viscous non-Newtonian inelastic as well as viscoelastic fluids of the memory integral viscoelastic K-BKZ (Kaye–Bernstein–Kearsley–Zapas) type. Their results show that the shear stress and the velocity in an arbitrary cross section, specifically a square cross section, corresponding to the constitutive equation of the purely viscous type (Eq. 2.5a) and corresponding to the viscoelastic fluid

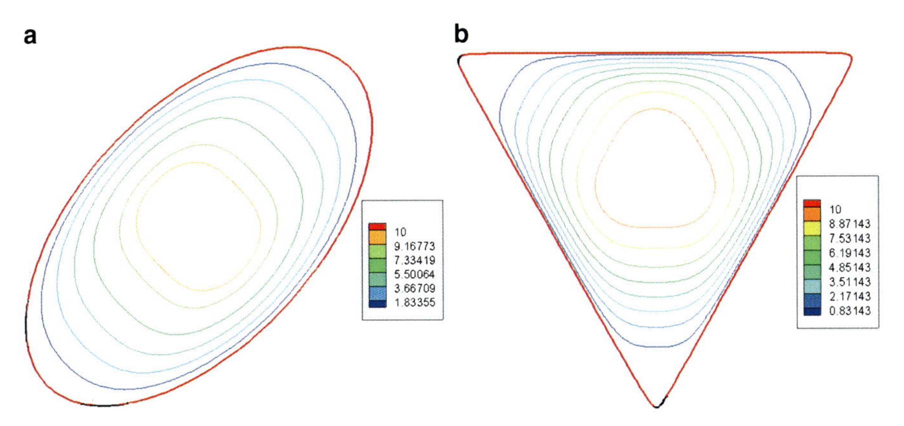

Fig. 2.1 Velocity level lines in an (**a**) elliptical contour ($n=2$, $\varepsilon=0.4$) and in a (**b**) triangular cross section ($n=3$, $\varepsilon=0.384$): $Re=200$, $We=0.3$, $\xi=0.3$. (n, ε) represent mapping parameters and Re, We and ξ stand for the Reynolds number, the Weissenberg number and the non-affine slippage constant, respectively. See Sect. 3.2 for a definition of the Weissenberg number and the slippage constant (reprinted from Siginer and Letelier [15] with permission)

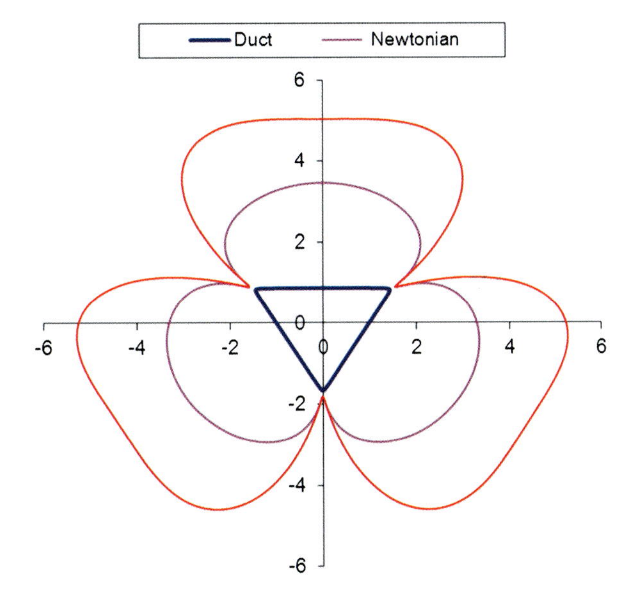

Fig. 2.2 Normalized wall shear stress τ_0 for Newtonian and viscoelastic fluids in a triangular duct $Re=200$: (**a**) Newtonian, (**b**) viscoelastic: $We=0.3$, $\xi=0.2$ (reprinted from Siginer and Letelier [15] with permission)

(Eq. 2.6) of the K-BKZ type, the Papanastasiou model [21] of the same zero shear viscosity μ_0, are quite different, thereby affirming that elasticity plays a significant role in determining the longitudinal flow field along the lines of the findings of Siginer and Letelier [15].

$$\mathbf{T} = -p\mathbf{1} + 2\eta(\mathbf{D})\mathbf{D} \tag{2.5a}$$

$$\eta(\mathbf{D}) = \eta_0 \left[1 + \left(\dot{\gamma} \eta_0 / \lambda \right)^a \right]^{-b} \tag{2.5b}$$

$$\mathbf{T}(t) = -p\mathbf{1} + \int_{-\infty}^{t} \Psi\left(\mathrm{I}_{\mathbf{C}^{-1}}, \mathrm{II}_{\mathbf{C}^{-1}}\right) \sum \frac{a_k}{\lambda_k} e^{-(t-\tau)/\lambda_k} \mathbf{C}_t^{-1}(\tau) \mathrm{d}\tau \tag{2.6}$$

In the above $\mathbf{T}$, p, $\mathbf{1}$, η, $\dot{\gamma}$, λ, Ψ, a and b represent the total stress tensor, the constitutively indeterminate part of the stress, the unit tensor, the shear rate dependent viscosity, the shear rate, the relaxation time, and a kinematic function of the first and second invariants $\mathrm{I}_{\mathbf{C}^{-1}}$ and $\mathrm{II}_{\mathbf{C}^{-1}}$ of the Finger tensor $\mathbf{C}_t^{-1}$, respectively. a and b in Eq. (2.5b) are curve fitting constants to be determined from experiments, and a_k and λ_k in Eq. (2.6) are the relaxation moduli and the relaxation times of the fluid.

The work of Siginer and Letelier [15] as well as of Ahmeda et al. [20] clearly demonstrates the effect of elasticity in computing the longitudinal field and the related components of the stress tensor. The conjecture that the longitudinal fields of viscoelastic fluids and purely viscous fluids of the power-law type are one and the same in tubes of arbitrary cross sections and can be predicted with good accuracy, if a modified Reynolds number of the Kozicki (Eq. $2.4)_2$ type is used, flows out of experimental measurements for rectangular cross sections which do show a 10 % difference from the predictions if the fluid is viscoelastic. But this much difference is in pace with the calculations of Siginer and Letelier [15] for arbitrary cross sections, that is, the influence of the fluid elasticity on the longitudinal field is of the order of 10 %. The fact that the use of the Kozicki Reynolds number may yield good results for viscoelastic fluids in rectangular cross sections may be due to a large extent to the empirical/phenomenological cross section-dependent constants $F(a)$ and $F(b)$ built in Eq. (2.4).

2.2 Turbulent Flow

An extensive review of the analytical and experimental studies for friction factor of Newtonian fluids for turbulent flow in non-circular ducts can be found in Bhatti and Shah [22].

2.2.1 *Viscoelastic Fluids: Pseudoplastic Behavior*

Predictions of the pressure drop of linear fluids in turbulent flow based on the assumption that the universal turbulent velocity profile in circular ducts is also valid for non-circular tubes are not accurate. This assumption is prompted by the observation that most of the momentum change in turbulent flow occurs in the laminar sublayer near the wall, and the velocity profile is relatively flat over the rest of the cross section. Hartnett et al. [23] found that experimental measurements with rectangular ducts of various aspect ratios are always larger than the predicted values with the square duct representing the worst case of departure; at Reynolds numbers higher than 10^4, the magnitude of the departure is about 12 % higher. The physical explanation lies with the secondary flow which accompanies the turbulent flow of linear fluids in non-circular ducts. Secondary flows provide a mechanism for the continuous transport of momentum from the central region of the tube toward the walls and in particular the corners thereby enhancing the velocity magnitudes there.

The fully developed turbulent friction factor for linear fluids flowing in rectangular cross sections and annuli can be predicted by formulas proposed by Jones [24] based on correlations of available experimental data. For instance, the friction factor f for rectangular ducts is given by

$$\frac{1}{\sqrt{f}} = 4.0 \log\left(Re_N^* \sqrt{f}\right) - 0.4,$$

$$Re_N^* = Re\,\Phi(\alpha^*), \qquad \Phi(\alpha^*) = \frac{1}{F(a) + F(b)}$$

where the Re is based on the hydraulic diameter D_h and $F(a)$ and $F(b)$ are the same tabulated functions which appear in Eq. (2.6). $\Phi(\alpha^*)$ can also be computed from an exact

$$\Phi(\alpha^*) = \frac{2(1+\alpha^*)^2}{3}\left\{1 - \frac{192}{\pi^5}\alpha^* \sum_{k=0}^{\infty} (2k+1)^{-5} \tanh\frac{(2k+1)\pi}{2\alpha^*}\right\},$$

or an approximate formula, accurate to 2 %,

$$\Phi(\alpha^*) = \frac{2}{3} + \frac{1}{8}\alpha^*(2 - \alpha^*).$$

A formula to determine the friction factor in the fully developed turbulent flow of generalized Newtonian fluids in circular pipes was proposed by Dodge and Metzner [25],

$$\frac{1}{\sqrt{f}} = \frac{4.0}{n^{0.75}} \log\left\{Re\, f^{\frac{(2-n)}{2}}\right\} - \frac{0.4}{n^{1.2}}. \tag{2.7}$$

The functional coefficients $\frac{4.0}{n^{0.75}}$ and $\frac{0.4}{n^{1.2}}$ in this equation are determined from correlations of experimental measurements of the pressure drop of aqueous Carbopol solutions and slurries of clay (Attagel) in the range $0.3 < n < 1$ for the power-law index. Rheological measurements taken on oscillatory rheometers clearly demonstrate that these are viscoelastic fluids, and yet they behave as purely viscous, that is, pseudoplastic fluids in turbulent flow in circular pipes without generating a drag-reducing effect. Dodge and Metzner [25] also proposed a universal velocity profile for generalized Newtonian fluids in turbulent tube flow divided into a laminar sublayer, a buffer layer, and a turbulent core region similar in overall structure to the Newtonian Karman–Nikuradse three-layer model where the interface between the buffer layer and the turbulent core region is at $y_2 = y_2^+$. The superscript $^+$indicates dimensionless entities represented in terms of wall coordinates normalized by the friction velocity $u\tau = \sqrt{\tau w/\rho}$ and kinematic viscosity ν, where τw is the shear stress at the wall and ρ is the fluid density.

$$u^+ = \left\{ \begin{array}{lll} (y^+)^{1/n} & y^+ < 5^n & \text{laminar sublayer} \\ \frac{5.0}{n}\ln y^+ - 3.05 & 5^n < y^+ < y_2^+ & \text{buffer layer} \\ \frac{2.78}{n}\ln y^+ + \frac{3.8}{n} & y_2^+ < y^+ & \text{turbulent core} \end{array} \right\},$$

As early as 1966 Kozicki et al. [26] suggested that the Dodge–Metzner equation (2.7) can be used to predict the pressure drop in turbulent flow in tubes of arbitrary cross section by substituting the hydraulic diameter for the pipe diameter. This suggestion was further modified by Kostic and Hartnett [27] who proposed to use Eq. (2.7) replacing Re with the Kozicki Reynolds number Re^* defined in Eq. (2.4). The validity of this approach was proven by Hartnett et al. [12] and Hartnett and Rao [28] who conducted experiments with Carbopol and Attagel solutions in square and rectangular ducts, respectively, for a wide range of power-law index values and for Kozicki Reynolds numbers in the range of $3{,}000 < Re^* < 50{,}000$.

2.2.2 *Drag-Reducing Viscoelastic Fluids*

2.2.2.1 Preliminaries

As it was pointed out in the previous paragraph, not all viscoelastic fluids are of the drag-reducing type. Carbopol solutions and Attagel clay slurries fall in that category, and their pressure drop in turbulent flows can be predicted following the guidelines outlined in the previous paragraph. However, the class of drag-reducing viscoelastic fluids such as those obtained by dissolving small amounts of long chain polymers in water or in another solvent behaves quite differently. Examples are polyacrylamide, Separan, polyethylene oxide, carboxymethylcellulose (CMC), and hydroxyethylcellulose (Natrosol). Reductions of more than 30 % in the frictional drag were observed with additive concentrations of less than 50 parts per million by weight (wppm). At very small concentrations of the order of 10 wppm, the measured apparent viscosity of these fluids does not show any deviation from the Newtonian viscosity, but the observed experimental friction factors are significantly lower than the Newtonian values. An increase in the concentration produces an increase in the apparent viscosity and in the characteristic relaxation time of the fluid accompanied by a corresponding decrease in the friction factor. The friction factor decreases with increasing concentration reaching an asymptotic value. Further increases in the concentration do not produce further decreases in the friction factor. This asymptotic behavior was first reported by Virk and his colleagues [29–31] who developed a comprehensive theory and correlation of drag-reducing activity based on their own work with pipe flow of polyethylene oxide solutions consistent with data obtained with other aqueous polymer solutions. It is interesting to note that pipe wall roughness is not a factor in drag reduction. Virk [32] reported that the onset of drag reduction in rough pipes occurred at the same wall shear stress as in smooth pipes, and the onset wall shear stress was essentially independent of polymer concentration and was unaffected by the flow regime (hydraulically smooth, transitional, or fully rough) during which onset occurs. The maximum drag reduction possible in the rough pipes was limited by an asymptote which was independent of polymeric parameters. Under asymptotic conditions, friction factors in all rough pipes identically obeyed the smooth pipe friction factor relation. Interestingly the maximum viscous sublayer thickness attained during drag reduction was approximately 2½ times that of Newtonian sublayer.

Experimental evidence is convincing that the extent of drag reduction is limited by a unique asymptote. The minimum drag asymptote in a circular geometry can be expressed both in implicit and explicit form, Virk et al. [31],

$$\frac{1}{\sqrt{f}} = 19.0 \log_{10}\left(Re\sqrt{f}\right) - 32.4,$$

$$f = 0.59\, Re^{-0.58}.$$

The latter expression for the asymptote is valid in the region $4 \times 10^3 \leq Re \leq 4 \times 10^4$. A slightly lower asymptote expressed in terms of the Reynolds number based on the apparent viscosity at the wall and which shows good agreement with experimental data in square and rectangular ducts has been proposed by Cho and Hartnett [13],

$$f = 0.2\, Re^{-0.48}.$$

The behavior of polyacrylamide solutions in water in circular pipes has been studied by Hartnett and Kwack [33]. They establish that the friction factor in turbulent flow for these fluids is a function of the Re and We numbers below a critical $We_{\mathrm{cr}} \sim 10$. For Weissenberg numbers We below this value the friction factor decreases with increasing We at a fixed Re reaching an asymptotic value at a $We \sim 10$. Beyond the critical value $We_{cr} = 10$, friction factor in fully established flow is a function of the Re only. The experiments were conducted in the range $10^4 < Re < 7 \times 10^4$. The critical Weissenberg number We_{cr} is not fixed for all Re and increases with increasing Re; for $Re = 10^4$ the value of $We_{cr} \sim 7$, whereas for $Re = 7 \times 10^4$ it assumes a numerical value larger than 10 about $We_{cr} \sim 11$. The same conclusions were reached by Ghajar and Azar [34] who experimented with other aqueous polymeric solutions, two different types of Separan. At a fixed Re for values of $We > We_{cr}$, the friction factor f is a constant. A critical We_{cr} is reached with decreasing We at a fixed Re. The friction factor increases with decreasing We for values of $We < We_{cr}$ gradually approaching the Newtonian value as the We approaches zero (see Sect. 3.5.2 for more on this topic).

2.2.2.2 Drag-Reducing Additives and Applications

Experimental pressure drop data for the turbulent flow of drag-reducing fluids in non-circular geometries are relatively rare in the literature with the exception of rectangular ducts. However, on the basis of the available evidence with rectangular tubes, the above results valid for circular cross sections can be cautiously extended to arbitrary non-circular geometries subject to future experimental verification. A very promising novel experimental technique based on nuclear magnetic resonance imaging (NMR) dubbed Rheo-NMR has been recently introduced, Schroeder and Jeffrey [35]. The real-time, non-invasive system does not require discreet tracers and allows the resolution of all velocity components, longitudinal and transversal, with equal ease. The technique has been known for some time going as far back as 1991 Callaghan [36]; however, its application to non-Newtonian flows occurred only recently.

The phenomenon of drag reduction, the dramatic decrease in pressure drop or equivalently the substantial reduction in the friction factor in turbulent tube flow when a minute amount of a high molecular weight polymer is added to a Newtonian fluid, thereby reducing wall shear stresses, was discovered by Toms [37] in 1948. However the first *recognized* drag reduction result is due to Mysels [38] who had determined earlier that the skin friction for gasoline in pipe flow was significantly reduced by the addition of an anionic surfactant. The process was patented by Mysels [39] in 1949. Since then, it has been found that many other types of additives, including wormy micelle-forming surfactants, bubbles, rigid fibers, and even solid spheres, lead to drag reduction in turbulent flow. Natural fibers such as wood pulp are preferable because they have a "low environmental load" and other advantages such as lack of degradation of the solution as polymeric additives do over time. Reduced energy loss in the turbulent pipe flow of wood pulp fiber suspensions in water was reported for the first time by Forrest and Grierson [40]. But their paper did not receive the attention it deserved and it went unnoticed. Synthetic fibers have the same advantage as natural fibers in terms of "degradation," but they require careful disposal as by their very nature they are chemicals and may contaminate rivers and soil when solutions are drained directly, very much like polymeric and surfactant solutions. Kubo and Ogata [41] recently suggested the use of bamboo fibers as drag-reducing agents. Wood pulp fibers produce a 20 % in energy savings as a ballpark figure at $Re \sim 2 \times 10^4$. Bamboo fibers generate the same energy savings, but cleared woods take much longer than bamboos to regenerate, thus the environmental advantage. Although fibers are chemically and mechanically stable in an aqueous environment, the use of fibers comes with a serious drawback as any type of fiber can cause plugging problems in pipelines due to the high concentration of fibers, as high as a few percent, required for drag reduction.

The applications in energy savings are widespread with intense research efforts in reducing energy losses in pipelines, drag reduction over ship and submarine hulls, and improving efficiency of district heating and cooling systems, in which chilled or heated water is generated at a central location in a city and pumped to buildings in the surrounding area. Drag-reducing additives greatly decrease system energy requirements, reducing pipe diameter or increasing flow rate. The first application of drag-reducing additive was in transport of crude oil in the Trans-Alaska (TAPS or Alyeska) Pipeline in 1979. The pipeline is 800 miles long with 48 in. diameter. Injecting a concentrated solution of a high molecular weight polymer downstream of pumping stations at homogeneous concentrations as low as 1 ppm increased crude throughput by up to 30 %, Burger et al. [42]. District heating or cooling systems provide or remove heat in buildings or a district by recirculating water heated or chilled at a central station. The water recirculation energy requirements make up about 15 % of the total energy for a district heating or cooling system. Surfactants can reduce pumping energy requirements by 50–70 %. The effectiveness depends on the kinds of additives used and the layout of the primary distribution system. Surfactant drag-reducing additives have been tested successfully in large-scale district heating systems in Denmark, Germany, and the Czech Republic and in district cooling systems in Japan with very significant savings in energy requirements. For instance, cationic surfactants in aqueous systems have been used in over 130 buildings throughout Japan and reduced pumping energy by 20–60 %. A novel application of surfactants is preventing flow-induced localized corrosion. Repeated impact of turbulent eddies on the wall causes intermittent stresses on the wall leading to mechanical damage to the surface material. Surfactants successfully suppress the formation of turbulent eddies near the wall.

2.2.3 *The Mechanism of Drag Reduction*

2.2.3.1 Experimental Findings

Experimental flow visualization techniques were developed in the late 1960s to probe and resolve the structure of the boundary layer in turbulent flows of homogeneous Newtonian fluids. The formation of low-speed streaks in the laminar sublayer in the region very near the wall and their interaction with the outer portions of the flow through a process of gradual lift-up, then sudden oscillation, bursting, and ejection into the main flow were discovered and investigated by Klein et al. [43]. Suspended solid colloidal size particles and high-speed motion picture camera moving with the flow were used by Corino and Brodkey [44] to investigate the viscous sublayer. They find that fluid elements are periodically ejected outward toward the centerline from a thin region adjacent to the sublayer. They identify a zone of high shear at the interface between the mean flow and the decelerated region that gives rise to the ejected elements. The interaction of the ejected elements with the mean flow in this high-shear region creates intense, chaotic velocity fluctuations, which are believed to be an important factor in the generation and maintenance of turbulence. Electrochemical techniques were used to measure the circumferential component of the velocity gradient in turbulent flow at the wall of a pipe by Sirkar and Hanratty [45], and hydrogen-bubble and hot-wire measurements with dye visualization were employed by Kim et al. [46] to investigate the boundary layer structure on a smooth-surface flat plate in a low-speed water flow to show that essentially all turbulence production occurs during intermittent *bursting* periods.

These developments encouraged Eckelman et al. [47] to look into the effect of drag-reducing additives on the production of turbulence in the viscous sublayer. The expertise gained and the methods developed for homogeneous Newtonian fluids were applied to turbulent flow of drag-reducing fluids starting in the early 1970s to examine the modification of the wall layer by polymeric additives and the turbulence production mechanism in the sublayer in the presence of additives, Donohue et al. [48], Fortuna and Hanratty [49]. The data of the former authors taken in the near-wall region of a fully developed two-dimensional channel flow showed that the spatially averaged bursting rate of low-speed streaks characteristic of the viscous sublayer decreased substantially suggesting that the dilute polymer solution decreases the production of turbulent kinetic energy in the near-wall region by inhibiting the formation of low-speed streaks. The authors explain this behavior, however, tentatively but perhaps for the first time, by high resistance to elongational strains and vortex stretching. Fortuna and Hanratty [49] conducted experimental studies of the influence of drag-reducing polymers on the time-averaged velocity gradient and on the two components of the fluctuating velocity gradient at the wall and concluded that the increase in drag reduction is accompanied by an increase of the size of longitudinally oriented eddies in the viscous sublayer. They speculate that polymer additives act to increase the viscous resistance in the transverse direction more than in the direction of the mean flow. Measurements made on a drag-reducing polymer solution in pipe flow using a novel type of laser Doppler meter developed by the author himself were reported in Rudd [50] with the conclusion that the polymer has very little effect upon the turbulent core of the flow, but thickens and stabilizes the viscous sublayer with turbulent intensity inside the sublayer unchanged; due to the thickening of the sublayer velocity, fluctuations just outside of it are greater. Dye visualization and motion picture techniques were used by Oldaker and Tiederman [51] to obtain a detailed description of the streak formation in the viscous sublayer. They find that the average transverse spacing of the streaks increases as the amount of drag reduction increases. The average streak spacing within the viscous sublayer is not a function of the distance from the wall in water flows and flows at lower levels of drag reduction. At high levels of drag reduction, the average spacing varies within the sublayer increasing as the wall is approached. Real-time hologram interferometry was used for flow visualization and turbulence measurements in the near-wall region in the spanwise direction and the direction normal to the wall by Achia and Thompson [52] to investigate the streaks and bursts that originate in the sublayer. Their data suggests a stabilized wall layer with less turbulence production. They also suggest the extensional viscosity of the dilute polymer solution as the most likely mechanism underlining this behavior. It was suggested by various researchers, among them Lumley [53] who argued that the buffer region is the area of importance in drag-reducing flows and that drag reduction can result if behavior in the sublayer and buffer layer differ. Reischman and Tiederman [54] examined this issue by taking velocity measurements in drag-reducing flows in a two-dimensional channel with a laser Doppler anemometer with solutions of polyacrylamide and polyethylene oxide producing drag reductions ranging from 24 to 41 %. They establish that the drag-reducing mean velocity profile can be divided into three zones: a viscous sublayer, a buffer or interactive region, and a logarithmic region. They find no evidence that the viscous sublayers of the drag-reducing channel flows are thicker than those in the solvent flows. In addition the normalized streamwise fluctuations are essentially the same in both the solvent and drag-reducing sublayers. They find that the changes caused by the polymer addition occur in the buffer region. The drag-reducing buffer region is thicker and the velocity profile in the outer flow region accommodates this buffer region thickening. The measurements of the streamwise velocity fluctuations also show that the polymer additives redistribute the primary turbulent activity over a broadened buffer region. The role of the linear sublayer and the buffer layer was further clarified by the experiments of Tiederman et al. [55] who determined that the presence of additives does not affect at all the spanwise spacing and bursting rate of the streaks in the linear sublayer, which stays the same as that of water by itself. But in the buffer zone, the dimensionless spanwise streak spacing increases and the average bursting rate decreases. The latter rate is larger than the former. Thus, the additives have a direct effect on the flow processes in the buffer region, and the linear sublayer appears to have a passive role in the interaction of the inner and outer portions of the turbulent wall layer. The work of McComb and Rabie [56, 57] is in general support of these findings as are the experimental results of Usui et al. [58]. The work of the latter authors, who measured turbulent characteristics of drag-reducing flow by laser Doppler velocimetry (LDV), suggests that polymer (aqueous solutions of polyethylene oxide) injection into a pipe flow caused a thickening of the buffer layer, enlargement of macroscale turbulent eddies, and suppression of fine-scale turbulent eddies. Luchik and Tiederman [59] reported that turbulence statistics are significantly modified even at very low concentrations of polymeric additive with approximately 30 % drag reduction at a concentration of 1–2 ppm. They also reported that the principal influence of the additives is to damp velocity fluctuations normal to the wall in the buffer region and that the average time between bursts increased for the drag-reducing flows as compared to the flow of water alone. The experiments of Walker and Tiederman [60] clarified the role of various terms in the Reynolds stress transport equation (see Sect. 6.1) in drag reduction. They use two-component laser velocimeter measurements in a fully developed turbulent channel flow to examine the effect of polymer injection on the Reynolds stresses and the production terms in the Reynolds stress transport equations. They determine that the production of the streamwise Reynolds normal stress was decreased, but production of the Reynolds shear stress was unchanged showing that the processes represented by pressure–strain correlation terms in the Reynolds stress transport equations may be directly affected by the polymer.

2.2.3.2 Theoretical and Numerical Findings

At the same time as these developments were taking place, the effects of polymer extensibility and relaxation on the onset and the extent of drag reduction were investigated through rheological studies of drag-reducing polymers, Metzner and Park [61], Seyer and Metzner [62], Metzner [63], Bewersdorff and Berman [64]. However in spite of all these efforts, the fundamental mechanism which controls drag reduction has been elusive. The phenomenon is perplexing even today, and a satisfactory description is not yet at hand. No comprehensive understanding of the interaction of rheology and turbulence has ever been within our reach, much less a single theory capturing the main features of the drag reduction phenomenon. However, substantial progress has been made in the last decade in understanding the basic underlying physics as indicated by the recent reviews, Graham [65] and Wang et al. [66]. Even though some details are emerging and are better understood, the full picture is still not quite clear. Two of the prominent and competing ideas to explain Toms phenomenon have been formulated in the 1970s and are still under discussion. These are comparison and correlation of characteristic time scales and characteristic length scales. Lumley [67, 68] indicates that even though conceptually it maybe possible that an interaction between polymer molecules and hydrodynamic structure may occur because of corresponding length scales determined from polymer chain size and turbulent eddy size, it is unlikely that may be the dominant mechanism as the polymer length scale is too small, the ratio of the polymer length to eddy size being of order 10^{-2}. Consequently it is more appropriate to choose a characteristic time as a correlating parameter. It is believed that it is more likely that Toms phenomenon arises because of the characteristic relaxation time of polymer molecules becoming coupled or aligning itself somehow with the time-dependent flow of turbulent eddies in such a way that energy is stored rather than being dissipated, Schowalter [69], the difficulty with this idea being the lack of reliable information on characteristic relaxation times to characterize polydisperse solutions. In the strain energy storage model proposed by Kohn [70], polymer molecules store energy when they are strained by high-shear stress near the wall and release it by relaxation when transported to the low-shear region at the core. The onset of drag reduction occurs when strain energy convection is comparable to energy diffusion. Another idea central to attempts to explain drag reduction is the high resistance to extension of polymer solutions. It is assumed that this high resistance to stretching may inhibit vortex formation and that this may inhibit the continuous stretching of vortex filaments and their constant entwining and mixing with other filaments, thereby leading to a reduction in energy dissipation. In the 1990s significant evidence was obtained through numerical simulations using a generalized Newtonian model for dilute solutions that an enhanced, preferably anisotropic, extensional viscosity leads to drag reduction in turbulent pipe flows, Toonder et al. [71], and turbulent channel flows, Orlandi [72]. Evidence from both numerical simulations and experiments points to reduced strength of the longitudinal vortices when long chain polymers are present in the flow. It is safe to say that up to this writing a vast majority of the proposed mechanisms of drag reduction are based on enhancing macromolecular resistance to extensional motions (high resistance to elongational strains) or elastic memory and relaxational effects. However, it is equally safe to state that the predictions of any theory put forth along these lines need improvement and direct uncontroversial evidence in support of any specific mechanism is still lacking.

The most promising tool that may have emerged starting with the mid-1990s is direct numerical simulation (DNS) due to advances made in numerical methods to resolve the structure of the boundary layer in the flow of dilute viscoelastic solutions and relate it to drag reduction. DNS has not only confirmed the established picture of wall-bounded turbulence but has also allowed a detailed evaluation of its structural and statistical features. The fully turbulent channel flow of a dilute polymer solution with the polymer chains modeled as finitely extensible elastic dumbbells based on the non-linear FENE-P model, which describes a polymer chain as two equal masses connected by a finitely extensible entropic spring (see Siginer [73] section 2.3.2.2), was investigated by Sureshkumar et al. [74] who used a direct numerical simulation (DNS) approach. The suitability of non-linear dumbbell models in describing the polymer conformation in *turbulent* flows was advocated earlier, Leal [75]. The simulation algorithm of Sureshkumar et al. [74] is based on a semi-implicit, time-splitting technique which uses spectral approximations in the spatial coordinates. The simulations show that the polymer induces several changes in the turbulent flow characteristics, all of them consistent with available experimental results. In particular, a decrease in the *rms* streamwise vorticity fluctuations and an increase in the average spacing between the streamwise streaks of low-speed fluid within the buffer layer were predicted (Fig. 2.3).

Since the quasi-streamwise vortices are known to play an important role in the production of turbulence, the observation of reduced *rms* streamwise vorticity seems to support a mechanism of drag reduction. These findings suggest a partial inhibition of turbulence generating events within the buffer layer by the macromolecules after the onset of drag reduction. They show that this inhibition is associated with an enhanced effective viscosity attributed to the extensional thickening properties of polymer solutions, previously proposed by Metzner, Lumley, and others. Using the simulation results obtained for different sets of parameter values which modify the relaxational and extensional properties of the model, a set of criteria for the onset of drag reduction is proposed. The existence of a critical Weissenberg number for the onset has been stipulated by several researchers in the past. For instance, based on their experiments Hershey and Zakin [76] proposed a transition Weissenberg number of $O(1)$ for fairly concentrated systems (>0.05 % by weight), but for a more dilute system (0.005 % by weight), a critical Weissenberg number $We_{cr} = 36$ was reported. Sureshkumar et al. [74] propose an onset Weissenberg number We_{cr} between 12.5 and 25 for dilute solutions based on simulations performed with the extensibility parameter $L = 10$ (L corresponds to the maximum

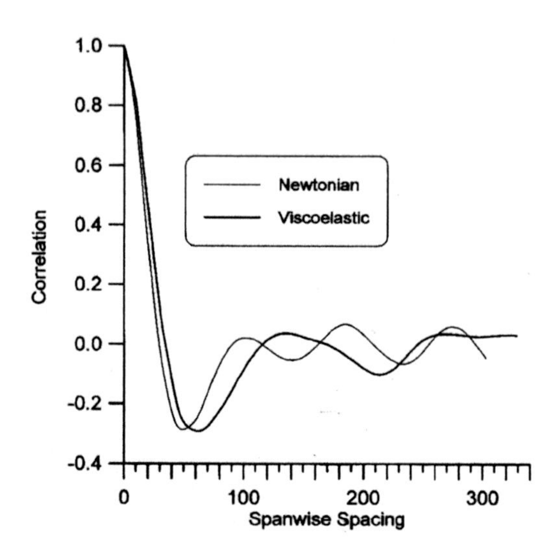

Fig. 2.3 The correlation of the streamwise fluctuating velocity as a function of spanwise spacing for both Newtonian and viscoelastic fluids (reprinted from Sureshkumar et al. [74] with permission)

value of the chain extension in the FENE-P model), which allows for sufficiently large extension of the polymer chains, contributing to an increased extensional viscosity. The Weissenberg number We_τ is defined as the product of the polymer relaxation time λ and a characteristic shear rate u_τ^2/ν_0 where ν_0 is the kinematic viscosity (diffusivity). u_τ is the friction velocity $u_\tau = \sqrt{\tau_w/\rho}$ where τ_w represents the shear stress at the wall and ρ is the density. Higher values of L indicate a high extensional viscosity and low values a low resistance to extensional strains. The maximum attainable extensional viscosity value is a monotonically increasing function of L for the FENE-P fluid; a high enough value of L is necessary to achieve significant drag reduction. Sureshkumar et al. [74] opt to use $L = 10$ in the computations after searching through a range of values for L. For instance, the mean flow velocity profiles obtained from the simulations performed for $L = 2$ at $We_\tau = 12.5$, 25, 37.5 and 50 are shown in Fig. 2.4 together with the Newtonian flow profile. All corresponding velocity profiles hardly deviate from Newtonian, and they all almost coincide indicating no detectable drag reduction for this low value of the elongational viscosity. For all the cases represented in Fig. 2.4, the root mean square velocity statistics are almost the same as the Newtonian flow results.

These observations are consistent with a mechanism for drag reduction requiring a significantly enhanced extensional viscosity in the bulk of the flow resulting in the suppression of the eddies which carry the Reynolds stress in the buffer layer, as proposed in the past by Lumley, Metzner, and others.

2.2.3.3 Factors Influencing the Effectiveness of Drag Reduction

Many factors have been identified that control the percent drag reduction with each drag-reducing additive.

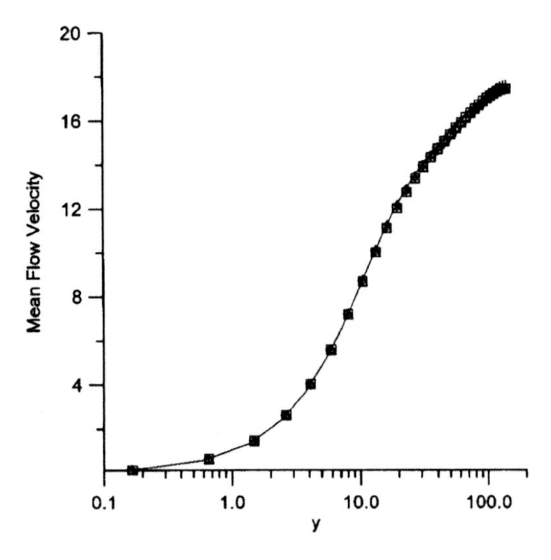

Fig. 2.4 The velocity $u(y)$ vs. y for the Newtonian (*solid line*) and four viscoelastic flows ($L = 2$) at $We = 12.5$ (*open diamond*), 25(+), 37.5 (*open square*), and 50 (*open triangle*) (reprinted from Sureshkumar et al. [74] with permission)

With fibers the effectiveness increases with fiber aspect ratio (length over diameter) and decreasing fiber diameter. Fibers mixed with polymers achieve drag reduction levels much higher than that for polymers or for fibers alone up to 95 %. An added advantage is that the polymer in a polymer fiber mixture is much less prone to degradation. With polymeric additives there are two well-established drag reduction types labeled type A and type B, Virk and Wagger [77]. Type A behavior occurs with very dilute solutions with which the onset of drag reduction at constant concentration only occurs in the fully turbulent regime above an onset Reynolds number. There is an onset shear stress and an onset shear rate. At Reynolds numbers less than the onset value, no drag reduction occurs. The onset may be due either to the stretching of the polymer molecules in the region of the flow field dominated by extension or to the entanglement of many molecules reaching the size of turbulent eddies. It is defined in terms of the onset shear stress or shear rate but not in terms of the onset Reynolds number because the latter is proportional to the 8/7th power of the pipe diameter, but the former is independent of it. The onset shear stress decreases with increasing molecular weight and radius of gyration of the polymer molecules, Virk [78]. Type B behavior happens with more concentrated polymer solutions. In this case the onset conditions can be reached at low Reynolds numbers. The laminar-turbulent transition is not observed, and laminar-like behavior is extended into an extended laminar region where drag reduction occurs. In both type A and type B behavior, drag reduction increases with flow rate until a critical wall shear stress is reached at which the rate of polymer degradation in the wall region exceeds the rate at which polymer is replenished in this region and drag reduction diminishes.

Other factors that influence the effectiveness of drag reduction are the molecular weight, the concentration effect, and the diameter effect. Polymers are generally ineffective for drag reduction if the molecular weight is less than 10^5, Hoyt

[79]. For a given concentration and Reynolds number, drag reduction increases with increasing average molecular weight, Virk [78]. Drag reduction at a fixed velocity increases with increasing polymer concentration until a saturation concentration is reached. Above this concentration, drag reduction starts falling off. The initial increase in drag reduction with increasing concentration is probably due to the increasing number of polymer molecules which cause the damping of a larger number of turbulent eddies. The decrease in drag reduction after the saturation concentration is caused by an increase in solution viscosity. With Newtonian fluids each Reynolds number in the turbulent region corresponds to a specific friction factor, which is independent of the tube diameter. In contrast at a given fixed Reynolds number, the same polymer solution in tubes of different diameter corresponds to different values of the friction factor. In general, the drag reduction observed in large pipes is smaller than that obtained in small pipe systems because of lower wall shear stresses and shear rates, Berman [80].

Surfactants are also effective drag reducers. In contrast to high polymers, their nanostructure can self-assemble after breakup by high shear, which makes them good candidates for use in recirculation systems containing high-shear pumps. Surfactant solutions are very sensitive to shear, which acts to induce reversible structural transformations in surfactant solutions manifesting themselves as shear thinning, shear-induced structures, shear-induced phase transitions (in shear bands), gelation, and flow instabilities, Fischer [81]. The mechanisms of these phenomena are not fully understood, although physical hypotheses and interpretation of the observed behavior have been formulated. For instance, it is thought that at low shear rates, surfactant solutions with rodlike or threadlike micelles usually act as Newtonian fluids because micelles rotate freely in the solution. At higher shear rates, micelles align themselves with the shearing direction causing shear thinning. Shear-induced structures occur at a critical shear rate at which the shear viscosity and elasticity show a sudden increase. These structures are orders of magnitude larger than the individual rodlike micelles, and the surfactant solution under these conditions behaves like a viscoelastic gel. The nature of shear-induced structures is not well understood in spite of considerable efforts. For instance, it is not known why it occurs with certain solutions and it does not with others.

In aqueous systems, when the critical surfactant concentration called the critical micelle concentration is reached, surfactant molecules gather into assemblies called micelles to minimize the hydrocarbon–water interface. When surfactant concentration is the same as or slightly above the critical micelle concentration, the shape of micelles is spherical or ellipsoidal. Micelles tend to form non-spherical shapes when the surfactant concentration reaches a second critical value. These shapes may be vesicles or disklike or may look like long cylinders called rodlike or wormlike micelles, which are generally considered excellent drag-reducing agents, Ohlendorf et al. [82]. The forces, which hold the surfactant molecules together in micelles, are much weaker than the primary chemical bonds of polymer molecules. But these forces persist even if the micelles encounter strong shear and breakup. They reform or self-assemble when the strong shear disappears, while polymer molecules cannot reform after mechanical degradation. The various factors that influence and govern the drag-reducing ability of surfactants such as micelle shape and size and the effect of both critical surfactant concentrations together with the type of surfactant have been the subject of intense research in the last few decades, in particular in view of the suitability of surfactants for the district heating and cooling applications. The drag-reducing properties and the temperature operating range of non-ionic surfactants, which do not carry any charge; anionic surfactants, which are the only known effective surfactant drag reducers in hydrocarbon media; cationic surfactants, which are not easily biodegradable; and zwitterionic surfactants, the molecules of which carry both positive and negative charges on different locations of the molecules have been the subject of intense investigations. There is limited evidence that zwitterionic/anionic mixtures containing up to 20 % anionic are most effective drag reducers.

Scale-up studies to predict drag reduction performance in large pipes of practical flow systems from small diameter measurements in the laboratory are important for applications, Gasljevic et al. [83]. The drawback of surfactant use for drag reduction is a significant reduction, always a little larger than drag reduction, in the heat transfer ability of the surfactant solution, which happens in tandem with drag reduction and poses a problem when heat exchangers are in the loop, Aguilar et al. [84]. The underlining physical mechanism behind this behavior is not understood. However, remedies to enhance the heat transfer ability of the drag-reducing surfactant solutions have been proposed all of them centering essentially on breaking up the micelle nanostructure before the fluid enters the heat exchanger.

Research in drag reduction has stayed very active over the last several decades given the implications in applications, in particular in energy savings. Reviews by Berman [80], Virk [78], and White and Hemmings [85] in the 1970s; Hoyt [79] in 1985; the books by McComb [86] and Gyr and Bewersdorff [87, 88] in the 1990s; and the recent reviews by Graham [65] in 2004, White and Mungal [89] in 2008, and Wang et al. [66] in 2011 give excellent overviews of the state of the science up to the date of publication. Drag reduction in multiphase flows is a topic which deserves attention in its own right. It has been surveyed by Manfield et al. [90] in 1999, but not since then. Extensive bibliographies were given by Nadolink and Haigh [91] up to 1995 (4,900 publications related to drag reduction between 1922 and 1994) and by Ge [92] in 2008.

Chapter 3
Transversal Flow Field

Abstract Early developments in the history of transversal flows, similarities with secondary transversal deformations associated with the simple shearing of solid materials in non-linear solid mechanics, as well as the analogy between the laminar flow of non-linear fluids and the turbulent flow of linear fluids in non-circular cross-sectional tubes together with constitutive criteria for the existence of secondary flows are commented on as well as the relatively recent research in secondary flows of dilute solutions in spanwise rotating pipes and channels and the related drag reduction in laminar flow. Recent findings on the fundamental aspects of transversal steady flows in straight tubes and the impact on industrial applications are reviewed emphasizing the importance of secondary flows in almost every industrial operation of significance involving constitutively non-linear fluids such as extrusion processes and viscous encapsulation. Only hydrodynamically and thermally developed flows are considered.

Keywords Secondary flows • Transversal velocity field • Straight tubes • Arbitrary cross-sections • Laminar flow of non-linear viscoelastic fluids • Analogy with turbulent flow of linear fluids • Analogy with transversal deformations of non-linear solids • Axial vorticity • Spanwise rotating pipes • Drag reduction in laminar flow • Industrial applications • Extrusion processes • Viscous encapsulation • Constitutive criteria • Existence of secondary flows • Weissenberg number • Deborah number

The transversal field in straight pipes is due to nascent second normal stress differences when the contour of the flow domain starts deviating slightly from the circular. Thus, it can be argued that in principle geometry is the root cause of the secondary field. In general, particles cannot follow straight pathlines in the pressure-gradient-driven flow of a viscoelastic fluid in a conduit with a cross section other than circular except when the apparent viscosity and the second normal stress function satisfy certain relationships independently of any constitutive assumptions.

D.A. Siginer, *Developments in the Flow of Complex Fluids in Tubes*,
DOI 10.1007/978-3-319-02426-4_3

3.1 Analogies

In addition to secondary flows, normal stress differences are the cause of non-linear phenomena such as the rod-climbing (Weissenberg effect) and die swell phenomena. It is interesting to note that the existence of the second normal stresses gives rise to similar corresponding phenomena in simple shearing of non-linear solid materials in elasticity and in shearing of granular materials referred to as Poynting effect and dilatancy, respectively. Another manifestation yet of normal stress differences, similar to secondary flows of non-linear fluids, that of the occurrence of secondary deformations in the shearing of non-linear elastic materials placed between eccentric cylinders has been investigated by Mollica and Rajagopal [93]. In a follow-up publication [94], they extend the analysis performed in [93] to the case of the shearing of a non-linear differential fluid, the fluid of grade three, between two eccentrically placed cylinders. The secondary flow streamlines and the transversal deformation pathlines determined in [94] and [93], respectively, bear a striking resemblance for driving forces of $O(1)$ with the same physical significance such as the longitudinal motion of the inner cylinder when the material in question, either solid or liquid, is placed between eccentric cylinders. For instance, a comparison of Fig. 4 in [93] and [94] which give the pathlines for the secondary displacement and the streamlines for secondary flow, respectively, due to the axial motion of the inner cylinder, and Fig. 6 in [93] and Fig. 7 in [94] which display the same in the case of an applied pressure gradient is very instructive (Figs. 3.1 and 3.2).

Another analogy with interesting constitutive implications, conjectured by Rivlin [95] and further explored by Speziale [96, 97] and Huang and Rajagopal [98], is with the emergence of secondary flows in turbulent motions of linear fluids in tubes of cross-sectional shapes slightly deviating from circular. Weak secondary flow structures of approximately 3 % of the strength of the primary flow in the flow of Newtonian fluids in non-circular tubes were observed by Nikuradse [99] as early as 1930. Analytical and computational evidence which shows that the circulation direction of the secondary vortices in rectangular cross sections due to elastic forces is opposite to that of the vortices prompted by turbulence exists in the literature; see, for example, Emery et al. [100] (also see Sect. 6.1 for a detailed exposition and further references). Inspired by Nikuradse's studies, Rivlin conjectured in [95] that there is a remarkable similarity between the laminar flow of non-linear fluids and the turbulent flow of linear fluids which led to the development of a class of constitutive formulations for turbulence, Speziale [96], Huang and Rajagopal [98]. Specifically, a one-to-one correspondence is established between the laminar flow of a non-Newtonian fluid of the rate type and the mean turbulent flow of a Newtonian fluid which leads to the second-order closure models of turbulence (see Sect. 6.1.1 for more on this). Interestingly depending on the specific modeling that is being done, different forms of the stress rate may or may not give rise to the generation of turbulent secondary flows as demonstrated by Speziale's work [96, 97]. This is an open issue in turbulence modeling and requires future work to bring clarity to the modeling of turbulent secondary flows.

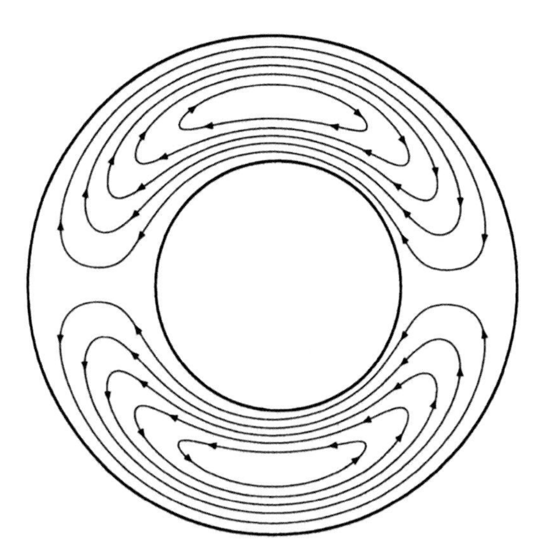

Fig. 3.1 Secondary flow streamlines due to the axial motion of the inner cylinder when the non-linear fluid is placed between two eccentric cylinders (reprinted from Mollica and Rajagopal [94] with permission) and pathlines for the secondary displacement when the deformation of the non-linear solid material placed between two eccentric cylinders is driven by the axial motion of the inner cylinder (reprinted from Mollica and Rajagopal [93] with permission). The graphs are indistinguishable from each other; consequently only one is reproduced here

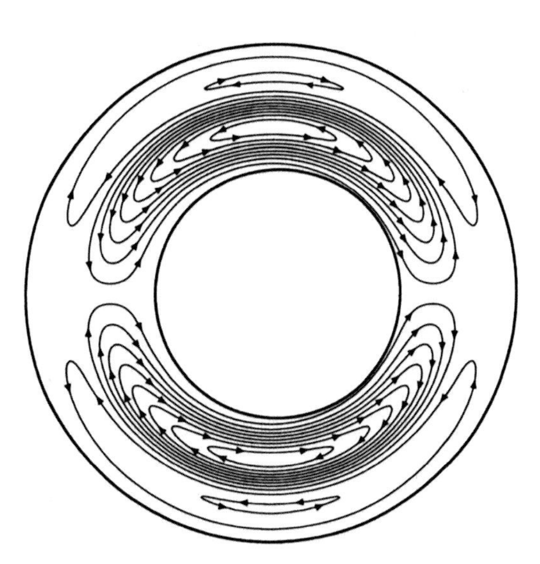

Fig. 3.2 Secondary flow streamlines of a non-linear fluid between two slightly eccentric cylinders due to an applied pressure gradient (reprinted from Mollica and Rajagopal [94] with permission) and pathlines for the secondary displacement when the deformation of the non-linear solid material placed between two eccentric cylinders is driven by an applied pressure gradient (reprinted from Mollica and Rajagopal [93] with permission). The graphs are indistinguishable from each other; consequently only one is reproduced here

3.2 Early Developments

Historically, Ericksen [101] was the first to recognize the existence of secondary flows of non-Newtonian fluids in straight pipes of non-circular cross section. He proved that steady rectilinear flow of a viscoelastic fluid in a straight pipe of non-circular cross section can be maintained if a suitable distribution of external body forces is applied to the fluid. In the absence of such body forces, he noted that the pathlines of a Reiner–Rivlin fluid could not be straight in a straight tube except when the tube cross section is circular or under very restricted conditions when the viscometric functions are not linearly independent and satisfy certain relationships.

The secondary flow of a Reiner–Rivlin fluid through pipes of elliptical cross section was predicted by Green and Rivlin [102] (Fig. 3.3). Their calculations implied that to a first approximation, secondary flows depend only on the second normal stress difference and that their strength was proportional to the driving pressure gradient raised to the fourth power with the implication that the intensity of secondary flows is much weaker than the primary longitudinal flow. That this is indeed so was further demonstrated by Langlois and Rivlin [103] and Rivlin [104] who extended the results of Green and Rivlin [102] to a more general class of non-Newtonian fluids, the class of order or equivalently grade fluids. Specifically, they use a fourth-order fluid, and through a perturbation analysis using the driving force as the perturbation parameter that is perturbing the rest state, they show that secondary flows arise at the fourth order. Further, they show that the ratio of the secondary flow stream function ψ to the primary longitudinal flow velocity w is a dimensionless number with physical significance, the ratio of elastic forces to viscous forces called Weissenberg number We raised to the third power,

$$\frac{\psi}{w} = \left(\frac{\lambda W}{L}\right)^3 = We^3,$$

with the implication that if the Weissenberg number is small, the secondary flow in the cross plane will be very weak. That this is indeed the case was further proven by Pipkin [105] who gave an ingenious qualitative discussion of the strength of secondary flows by means of dimensional analysis and symmetry considerations to reach the same conclusions. λ, W, and L represent the relaxation time, a

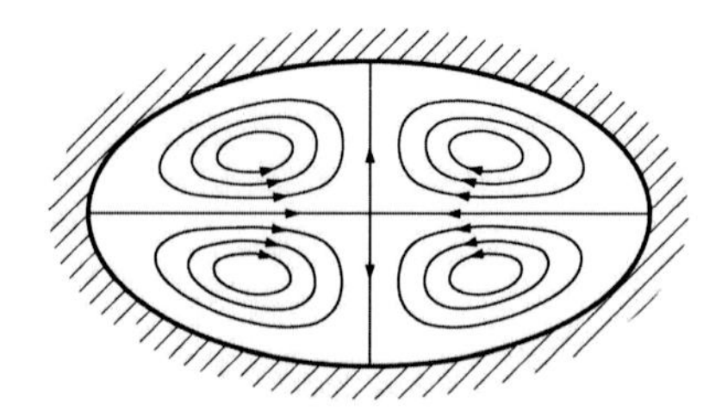

Fig. 3.3 Isothermal secondary flow in an elliptical tube (reprinted from Green and Rivlin [102] with permission)

characteristic velocity, and characteristic length, respectively. Wheeler and Wissler [106] are the first investigators on record who demonstrated the existence of secondary flows in square cross-sectional tubes. Like Green and Rivlin, they used the Reiner–Rivlin equation to define the behavior of the fluid. They used a finite difference technique with a power-law viscosity and a constant second normal stress. They also offered for the first time some sample experimental evidence of the existence of secondary flows. Up to that time, experimental evidence was lacking to corroborate theoretical findings. About the same time in 1965, Giesekus [107] conducted an experimental investigation with 5 % aqueous solution of very high molecular weight polyacrylamide and observed secondary flows in an elliptical cross section which agreed qualitatively with theoretical predictions. In particular, he observed that fluid particles along the long axis of the ellipse move toward the center of the channel and thus feed the secondary motions and particles move along the small axis toward the wall. More comprehensive experiments with various polymer melts were conducted soon after by Semjonow [108] who verified and further extended the observations of Giesekus [107].

3.3 Perturbation Approach: Non-linear Fluid and Solid Mechanics

3.3.1 Similarities and Constraints on Constitutive Formulations

Fosdick and Kao [109] had the insight of recognizing the implications of the counterpart in non-linear elasticity of the ideas developed by Ericksen [101] for non-linear fluids. The former authors recognized that the field equations governing the simple shearing of non-linear fluids have the same structure as those defining the rectilinear shearing of non-linearly elastic solids and that as a consequence there would be transversal deformations that accompany rectilinear shear unless the material functions characterizing the non-linear solid satisfy certain relationships.

An important difference in perturbation studies of secondary flows in fluid mechanics as well as of transversal deformations in non-linear solid mechanics is the perturbation parameter employed which may lead to results with vastly different physical significance. For instance, Fosdick and Kao [109] investigate the problem of the secondary deformations due to the axial shearing of a non-linear solid located in the annular region between two eccentrically placed cylinders using the driving force, that is, the applied shear, as the perturbation parameter to determine that transversal deformations arise for the first time at the fourth order of the algorithm. They observe that purely rectilinear shear is possible in a Mooney–Rivlin material and pick a model which represents the Cauchy stress $\mathbf{T}$ at any point $\mathbf{x}$ of the body in terms of the left Cauchy–Green stretch tensor $\mathbf{B}$,

$$\mathbf{T} = -p\mathbf{1} + \mu\mathbf{B} + \nu\mathbf{B}^2$$

with conditions provided on the material functions μ and ν to ensure that the material will necessarily undergo secondary deformations when sheared axially. In particular, the shear modulus is required to be shear dependent. Of course in a Mooney–Rivlin material, shear modulus is a constant. p and $\mathbf{1}$ represent the constitutively indeterminate part of the total stress and the unit tensor, respectively, with the constitutive constants μ and ν. The equivalent in non-linear fluid mechanics of this type of restrictions on material functions to produce secondary flows was conjectured by Oldroyd [110] and derived rigorously by Fosdick and Serrin [111] as we shall see later. The pioneering work of Stone [112] who was the first to examine closely the conditions under which the rectilinear motion of a plastic solid or a non-Newtonian fluid is possible in tubes of non-circular cross section deserves special mention as well as the work of Pipkin and Rivlin [113] who demonstrated that secondary flows constitute a non-linear normal stress effect. Earlier, Criminale et al. [114] had shown that if secondary flows do not occur in an elliptical tube, then they will not occur in any tube of arbitrary cross section.

3.3.2 *Types of Perturbation*

The technique of Fosdick and Kao [109] is equivalent to perturbing the rest state. This is the same technique as that of Truesdell and Noll [115] who provided a formal systematic perturbation approach perturbing the rest state implicitly assuming that the driving force is small. If the displacement field in solids or the velocity field in fluids is denoted by $\mathbf{u}$ at any point $\mathbf{x}$ in the body and the perturbation parameter is ε, the algorithm is described by $\mathbf{u} = \sum_{n=1}^{\infty} \varepsilon^n \mathbf{u}_n(\mathbf{x})$. Truesdell and Noll [115] assume without proof that the behavior of a general simple fluid can be approximately predicted by asymptotic approximations obtained through retardation (see Siginer [73] for more on this subject). Proving the convergence of the hierarchical asymptotic approximations stemming from retarding the motion would be most daunting if at all possible. No rigorous proof of convergence has been constructed to this day. Nevertheless, considerable light has been shed on the problem by Dunn and Rajagopal [116] who showed in 1995 that the retardation procedure does not provide a hierarchy of models as it was generally believed up to that time.

In contrast to the technique of Fosdick and Kao [109], Mollica and Rajagopal [93] use the deviation from concentricity as the perturbation parameter in studying the same problem as Fosdick and Kao [109], with an equivalent stress representation expressed in terms of the stored energy function $\sigma(\mathrm{I}_\mathbf{B}, \mathrm{II}_\mathbf{B})$, where $\mathrm{I}_\mathbf{B}$ and $\mathrm{II}_\mathbf{B}$ represent the first and second invariants of the Cauchy–Green stretch tensor $\mathbf{B}$ and same restrictions on the material parameters δ_1, δ_2, n and b

$$\mathbf{T} = -p\mathbf{1} + 2\frac{\partial \sigma}{\partial \mathrm{I}_{\mathbf{B}}}\mathbf{B} - 2\frac{\partial \sigma}{\partial \mathrm{II}_{\mathbf{B}}}\mathbf{B}^{-1},$$

$$\mathrm{I}_{\mathbf{B}} = tr\mathbf{B}, \qquad \mathrm{II}_{\mathbf{B}} = \frac{1}{2}\left[(tr\mathbf{B})^2 - tr\mathbf{B}^2\right]$$

$$\sigma(\mathrm{I}_{\mathbf{B}}, \mathrm{II}_{\mathbf{B}}) = \frac{\delta_1}{2b}\left\{\left[1 + \frac{b}{n}(\mathrm{I}_{\mathbf{B}} - 3)\right]^n - 1\right\} - \frac{\delta_2}{2}(\mathrm{II}_{\mathbf{B}} - 3) \tag{3.1}$$

$$\delta_1 > 0, \qquad \delta_2 < 0, \qquad 1 > n > \frac{1}{2}, \qquad b > 0$$

allowing for large driving forces but for small eccentricities to find that transversal deformations arise at the first order rather than at the fourth order. This is not a perturbation of the rest state and allows for a large primary field in the absence of eccentricity. It is a perturbation of the rectilinear deformation field which exists when eccentricity is zero. Their technique is not limited to small primary deformations in the axial direction in the case of a solid and is not limited to slow flows in the longitudinal direction in the case of non-linear fluids in the sense that the driving force is not required to be small. The algorithm is given by

$$\mathbf{u} = \mathbf{u}_0(\mathbf{x}) + \sum_{n=1}^{\infty} \varepsilon^n \mathbf{u}_n(\mathbf{x}).$$

Thus when ε is zero, there is a non-null purely axial solution, that is, when the eccentricity is zero, there is an axial field, not small, between the concentric cylinders driven either by an applied pressure gradient or the axial motion of either or both boundary.

Mollica and Rajagopal [94] solve the problem of the motion between slightly eccentric cylinders of a simple fluid of the differential type whose Cauchy stress $\mathbf{T}$ is represented by

$$\begin{aligned}\mathbf{T} = -p\mathbf{1} + \mu\mathbf{A}_1 + \alpha_1\mathbf{A}_2 + \alpha_2\mathbf{A}_1^2 + \beta_1\mathbf{A}_3 + \beta_2(\mathbf{A}_1\mathbf{A}_2 + \mathbf{A}_2\mathbf{A}_1) \\ + \beta_3\left(tr\mathbf{A}_1^2\right)\mathbf{A}_1\end{aligned} \tag{3.2}$$

where $\mathbf{A}_i$ are the Rivlin–Ericksen tensors [117] defined in terms of the rate of deformation tensor $\mathbf{D}$ through the recursive relationship (see also Siginer [73]) and $\mu, \alpha_1, \alpha_2, \beta_1, \beta_2, \beta_3$ are constitutive constants.

$$\mathbf{L} = \mathrm{grad}\,\mathbf{u}, \qquad 2\mathbf{D} = \mathbf{L} + \mathbf{L}^{\mathrm{T}}.$$

$$\mathbf{A}_1 = 2\mathbf{D}, \qquad \mathbf{A}_{n+1} = \mathbf{A}_n + \mathbf{L}^{\mathrm{T}}\mathbf{A}_n + \mathbf{A}_n\mathbf{L}$$

This model comes out of the retardation procedure in the sense of Coleman and Noll [118] (see also Siginer [73], section 2.3.3), that is, it may be conceived of as an approximation to a simple fluid. But Mollica and Rajagopal [94] consider Eq. (3.2) to be an exact model, fluid of grade three, in its own right in the sense described by Fosdick and Rajagopal [119], not an approximation, which unlike the retardation approximation places restrictions on the material coefficients via thermodynamical considerations. The Clausius–Duhem inequality is required to hold, and the Helmholtz free energy is constrained to be a minimum when the fluid is locally at rest, in particular $\beta_1 = \beta_2 = 0$ and $\mu \geq 0$, $\alpha_1 \geq 0$, $\beta_3 \geq 0$, $|\alpha_1 + \alpha_2| \leq \sqrt{24\mu\beta_3}$ due to these thermodynamical requirements reducing the thermodynamics compliant form of Eq. (3.2) to

$$\mathbf{T} = -p\mathbf{1} + \mu\mathbf{A}_1 + \alpha_1\mathbf{A}_2 + \alpha_2\mathbf{A}_1^2 + \beta_3\left(tr\mathbf{A}_1^2\right)\mathbf{A}_1$$

It is pertinent to remark at this point that available experimental data indicates that $\alpha_1 < 0$ for the working fluids in the experiments, which poses a dilemma as it contradicts the thermodynamically dictated restrictions. One may than conjecture that the fluids in the experiments are not fluids of grade three or they abide by a more encompassing constitutive equation of which the fluid of grade three is the truncated form. There has been no definitive resolution to this day to these conflicting experimental and analytical results. However, some light has been shed on the issue by the seminal paper of Müller and Wilmanski [120]. Thermodynamics places the restrictions $\mu > 0, \alpha_1 > 0, \alpha_1 + \alpha_2 = 0$ on the material constants of the second-order fluid taken as an exact model in its own right for compatibility with the second law, Dunn and Fosdick [121]. However, experiments indicate that $\mu > 0, \alpha_1 < 0, \alpha_1 + \alpha_2 \neq 0$. Further, Dunn and Fosdick [121] showed that disregarding thermodynamics and assuming that $\alpha_1 < 0$ lead to instabilities and in quite arbitrary flows instability and boundedness are unavoidable. That the instabilities persist even if $\mu > 0, \alpha_1 < 0, \alpha_1 + \alpha_2 \neq 0$ is assumed was shown by Fosdick and Rajagopal [119]. In addition, Joseph [122] has shown that for the second-order fluid and indeed for fluids of arbitrary higher order, the rest state is unstable. Müller and Wilmanski [120] rederive the second-order fluid model within the framework of the *irreversible extended thermodynamics* introduced by Müller [123] and find that α_1 is allowed to be negative. They also find that there are no instabilities, and the speeds of the shear waves determined by the viscosity and the normal stress coefficients are finite. The same holds for the family of all *rate-type* equations, sometimes called equations of the Maxwellian type, of which the second-order fluid is part of. Extended thermodynamics puts different restrictions on the material parameters of *rate-type* equations. The idea behind *irreversible extended thermodynamics* is to take the stress and the heat flux as *independent* variables. Each one of them is governed by a balance equation. As balance equations are not conservation equations, they include fluxes. The immediate consequence is that the classical constitutive equations, the Navier–Stokes theory and the Fourier's law, are *not any longer constitutive equations* in their own right but rather arise at the first order of the Maxwellian iteration process, Müller, [123], Müller and Wilmanski [120] (also see Siginer [73], section 2). Following the lead of Müller and Wilmanski, the Reiner–Rivlin and

Rivlin–Ericksen second-order fluids are derived by Lebon and Cloot [124] in the framework of extended thermodynamics, and the resulting equations with the attached thermodynamics restrictions imposed by irreversible extended thermodynamics were used to analytically solve the problem of the Marangoni convection in a thin horizontal layer of a non-Newtonian fluid subjected to a temperature gradient in microgravity. Although it is not directly related to non-Newtonian fluid mechanics, we note that Depireux and Lebon [125] studied non-Fickian diffusion in a two-component mixture at uniform temperature in the framework of *extended irreversible thermodynamics.* The idea here is to elevate the dissipative diffusion flux to the status of independent variable and to derive the evolution equation for the diffusion flux to show that in the linear regime, the classical Fick's laws are recovered. The results in the non-linear regime lead to non-Fickian diffusion laws different from the well-known Fick's laws.

In addition to the restrictions imposed by thermodynamics, material parameters are to satisfy certain restrictions in the sense of Fosdick and Serrin [111] (see below) to ensure that secondary flows take place. In a way, the most important of these read $2\alpha_1 + \alpha_2 \neq \text{const.}\ f(\kappa^2)$, $f(\kappa^2) = \mu + 2\beta_3\kappa^2$, that is, $2\alpha_1 + \alpha_2$ is not to be equal to a constant value of $f(\kappa)$ for the secondary flows to take place. Violation of this condition or for that matter any one of the other conditions required by the theorem of Fosdick and Serrin [111] will lead to a rectilinear flow between slightly eccentric cylinders either pressure driven or driven from the boundary. A parallel example in non-linear solid mechanics of this type of restrictions is the problem studied by Mollica and Rajagopal [93]. The restrictions (Eq. 3.1) ensure that secondary deformations take place. If $n = 1$ or $b = 0$ or $\delta_1 = 0$ or $\delta_2 = 0$, the model will only predict purely axial shear. In fact for $n = 1$, the model reduces to the classical Mooney–Rivlin material.

A third way to study these constitutively non-linear problems is to use as the perturbation parameter a material parameter which captures the deviation from the Newtonian behavior as in the problem investigated by Jones [126] or a material parameter which captures most the physics of the problem even though there may be other material parameters as in the problems studied by Dodson et al. [127] and Townsend et al. [128]. Jones [126] studied the secondary flow of a generalized Reiner–Rivlin fluid between eccentric cylinders moving relative to each other. The single constitutive parameter, which captures the deviation from Newtonian behavior, is used as the perturbation parameter. His approach suffers neither from small driving forces (slow flows) nor small eccentricities. Even though the approach is not limited to slow flows, it suffers from low intensities for the strength of secondary flows as the deviation from linear behavior is small.

In contrast to the apparent viscosity and the first normal stress difference, determination of the second normal stress difference was and remains even today a challenge. At the time Dodson et al. [127] conducted their investigation, even the sign of the second normal stress difference was controversial. It was believed that the sign of N_2 was negative for some liquids and positive for others and that the sign of N_2 may even change with increasing shear. The pioneering investigations of Dodson et al. [127] and Townsend et al. [128] sought to determine the sign and magnitude of N_2 from a study of the direction and strength of secondary flows in rectangular and square straight pipes. They adopted a variable coefficient constitutive equation exact for viscometric flows, the CEF (Criminale-Ericksen-Filbey) equation derived by Criminale et al. [114] (see also Siginer [73], section 2.4.1),

$$\mathbf{S} = 2\eta\mathbf{D} - \Psi_1 \overset{\nabla}{\mathbf{D}} + 4\Psi_2\mathbf{D}^2$$

$$\eta = \frac{S_{12}}{D_{12}}, \qquad \Psi_1 = \frac{N_1}{D_{12}^2}, \qquad \Psi_2 = \frac{N_2}{D_{12}^2}$$

$$\overset{\nabla}{\mathbf{D}} = \overset{\circ}{\mathbf{D}} - 2\mathbf{D}^2$$

$$\mathbf{S} = 2\eta\mathbf{D} - \Psi_1 \overset{\circ}{\mathbf{D}} + (2\Psi_1 + 4\Psi_2)\mathbf{D}^2$$

assuming that it would be approximately valid for the nearly viscometric flow in the tube as the strength of the secondary flow is likely to be weak and consequently the resulting helical particle pathlines forming the vortices in the cross plane will be stretched out with a very long directrix. $\mathbf{S}$, $\mathbf{D}$, η, Ψ_1, Ψ_2, $\overset{\nabla}{(\bullet)}$, and $\overset{\circ}{(\bullet)}$ in the above stand for the total stress tensor, the rate of deformation tensor, the viscosity, the phenomenological coefficients (the first and second normal stress coefficients Ψ_1 and Ψ_2 with N_1 and N_2 representing the first and second normal stresses), the upper convected derivative, and the corotational time derivative, respectively [for definitions of the upper convected derivative and the corotational time derivative, see Appendix A and also Eq. (3.4) for the latter]. It is worthwhile to remark that the second-order fluid model is the same as the CEF equation in structure except that second-order fluid model has constant coefficients. The latter applies to any flow that is *slow and slowly varying* and the former to any *steady shearing* flow. Dodson et al. [127] assume that the coefficients of the CEF equation are functions of the second invariant of the flow thus independent of the shear rate and consequently constants and use the constant second normal stress difference as the perturbation parameter. The form of the apparent viscosity and the first normal stress difference are not restricted in their computations. They reach the conclusion that the direction of the secondary flow streamlines is reversed if the sign of N_2 is changed. In addition, they carry out flow rate and secondary flow visualization experiments and from a comparison of the observed direction of the secondary flows and their theoretical predictions based on a first-order solution in $O(N_2)$ to conclude that the sign of the second normal stress difference is positive. We now know that is not the case. That this was not the case was suspected by Townsend et al. [128] who conducted experiments with six test fluids in square sectional tubes and an extension of the theoretical analysis of Dodson et al. [127] to conclude that the direction of the streamlines is not indicative of the sign of N_2. The difference between the analyses of these two investigations is that Townsend et al. [128] did not identify N_2 as the perturbation parameter and did not take it to be a constant. As a consequence, N_2 was not restricted and the derivatives of N_2 were included in the equations. That had a marked effect on the conclusions reached in that no reversal of the secondary flow streamline direction could be predicted if the sign of N_2 is changed.

3.4 Criteria for the Existence of Secondary Flows

Although Stone [112], Criminale et al. [114], Pipkin and Rivlin [113], and Oldroyd [110], in that order, laid out the groundwork, a mathematically rigorous and general theorem concerning the existence of secondary flows in straight tubes of arbitrary cross section was proved by Fosdick and Serrin [111]. Earlier the structure of secondary flows of simple fluids in tubes of any arbitrary cross section was determined by Truesdell and Noll in their monograph [115] through a systematic study via a perturbation approach. Fosdick and Serrin [111] showed that unless the cross section of the tube is circular or an annulus between two concentric circles, a simple fluid cannot undergo a steady rectilinear motion in a straight tube whose cross section is a bounded and connected set assuming that the material functions satisfy appropriate analyticity and monotonicity conditions and that they are not proportional for small shear rates. The precise theorem they proved reads "Suppose an incompressible simple fluid moves rectilinearly and steadily in a fixed straight tube whose cross section is a bounded and connected open set. Assume also that the adherence condition or the slip condition is satisfied at the tube wall. Under these circumstances if the material functions φ and μ related to the viscometric functions of the fluid and dependent on the shear rate satisfy appropriate analyticity and monotonicity conditions, and if φ is not proportional to μ for small shear rates, then the cross section of the tube must be either circular or the annulus between two concentric cylinders." It is interesting to note that shortly after Fosdick and Serrin published their results, McLeod [129] proved two theorems in his study of the overdetermined systems, one of which includes the results obtained by Fosdick and Serrin almost as an immediate consequence.

The conditions for steady rectilinear flow of a viscoelastic fluid to take place in tubes of arbitrary cross section were previously conjectured by Oldroyd [110] through momentum considerations. Rectilinear flow can occur in a straight tube of arbitrary cross section independently of any constitutive assumptions if any one of the following three conditions is met:

(a) If the second normal stress difference is zero: $N_2\left(\dot{\gamma}\right) = (T_{rr} - T_{\theta\theta})$ where (r) and (θ) represent the radial and azimuthal coordinates and $\mathbf{T}$ and $\dot{\gamma}$ stand for the total stress tensor and the shear rate, respectively

(b) If both the apparent viscosity $\eta\left(\dot{\gamma}\right)$ and the second normal stress coefficient $\Psi_2\left(\dot{\gamma}\right)$ are constant: $N_2\left(\dot{\gamma}\right) \neq 0$, $\eta\left(\dot{\gamma}\right) = \text{Const.}$, $\Psi_2\left(\dot{\gamma}\right) = \text{Const.}$, $\Psi_2\left(\dot{\gamma}\right) = N_2\left(\dot{\gamma}\right)/\dot{\gamma}$

(c) If the apparent viscosity and the second normal stress coefficient are proportional: $\eta\left(\dot{\gamma}\right) = m\Psi_2\left(\dot{\gamma}\right)$ where m is a coefficient of proportionality.

If none of the above conditions is met, a secondary flow occurs in the transversal plane which causes the particles to follow a spiraling path down the tube. These are hierarchical conditions in the sense that it is not enough that the second normal stress difference is not zero for the particular constitutive equation in use to predict secondary flows. In other words, a non-zero second normal stress difference is a necessary but not a sufficient condition for the existence of secondary flows. It may very well be that the apparent viscosity and the second normal stress coefficient predicted by the constitutive equation are both constant. If that is the case, the constitutive equation will not predict any secondary flows even though the second normal stress difference is non-zero. It is worthwhile to note that in the latter case, the magnitude of the departure of the second normal stress coefficient from a constant multiple of the apparent shear viscosity determines the strength of the secondary flow. Thus, fluids which obey the upper convected Maxwell or the Oldroyd-B models and some versions of the Phan-Thien–Tanner model will not develop secondary flows and will display rectilinear particle pathlines when flowing in straight tubes of cross section other than circular. In the case of Maxwell and Oldroyd-B constitutive structures the second normal stress difference $N_2\left(\dot{\gamma}\right)$ is zero. Some versions of the Phan-Thien–Tanner model produce a constant second normal stress coefficient $\Psi_2\left(\dot{\gamma}\right) = N_2\left(\dot{\gamma}\right)/\dot{\gamma}$ and constant apparent viscosity $\eta\left(\dot{\gamma}\right)$; other versions predict $\Psi_2\left(\dot{\gamma}\right) = m\eta\left(\dot{\gamma}\right)$ where m is a coefficient of proportionality. These secondary flows are independent of end effects as they occur when the velocity field and the driving pressure gradient for instance are independent of the axial coordinate. A compendium of the most popular constitutive equations is given in Appendix A.

Comments at this point on two popular constitutive equations are in order in the light of the relationships the rheological functions of the fluid must satisfy in order for the secondary flows to exist in the computational domain. If we consider the fluid of order two or equivalently the Rivlin–Ericksen fluid of grade two, with constant coefficients (material parameters) condition (b) above is met, and no secondary flows may exist in the computational domain.

$$S_{kl} = \mu \mathrm{A}_{kl}^{(1)} + \frac{1}{4}\alpha_1 \mathrm{A}_{km}^{(1)} \mathrm{A}_{ml}^{(1)} + \alpha_2 \mathrm{A}_{kl}^{(2)}$$

On the other hand if constant μ is replaced by a point function in the field, an apparent viscosity given by $\eta = f(tr\mathbf{D}^2)$, secondary flows will be predicted in the computational domain. Whether these will be representative of the actual flows observed is a completely different matter. A second example is the single mode corotational Maxwell fluid,

$$\mathbf{T} = -p\mathbf{1} + \mathbf{S}$$

$$S_{kl} + \lambda \overset{\circ}{S}_{kl} = 2\mu D_{kl} \tag{3.3}$$

$$\overset{\circ}{\mathbf{S}} = \overset{\bullet}{\mathbf{S}} - \left(\nabla \mathbf{u}^{\mathrm{T}} - \mathbf{D}\right)\mathbf{S} - \mathbf{S}\left(\nabla \mathbf{u}^{\mathrm{T}} - \mathbf{D}\right)^{\mathrm{T}}, \tag{3.4}$$

where $\overset{\bullet}{\mathbf{S}}$ and $\overset{\circ}{\mathbf{S}}$ represent the material derivative and the corotational Jaumann derivative of the extra-stress $\mathbf{S}$ written in terms of the velocity $\mathbf{u}$ and the rate of deformation tensor $\mathbf{D}$ and λ is the single relaxation time. No secondary flows will be predicted for constant μ because condition (c) above is met,

$$\left\{ \begin{aligned} \eta\left(\overset{\bullet}{\gamma}\right) &= \frac{\mu}{1 + \lambda^2 \overset{\bullet}{\gamma}} \\ \Psi_2\left(\overset{\bullet}{\gamma}\right) &= -\frac{\lambda\mu}{1 + \lambda^2 \overset{\bullet}{\gamma}} \end{aligned} \right\} : \Psi_2\left(\overset{\bullet}{\gamma}\right) = -\lambda\, \eta\left(\overset{\bullet}{\gamma}\right).$$

On the other hand, this equation would predict secondary flows in the computational domain if the apparent viscosity were given by $\eta = f(tr\mathbf{D}^2)$ as none of the above conditions would have been met. A realistic dependence of the apparent viscosity on the second invariant of $\mathbf{D}$ can be thought of, for instance, as a truncated power law

$$\eta = \left\{ \begin{array}{ll} \mu & \text{when } \mathrm{II}_{\mathbf{D}} \le \mathrm{II}_{\mathbf{D}}^0 \\ \mu\left(\mathrm{II}_{\mathbf{D}}/\mathrm{II}_{\mathbf{D}}^0\right)^{m-1} & \text{when } \mathrm{II}_{\mathbf{D}} > \mathrm{II}_{\mathbf{D}}^0 \end{array} \right\}, \quad \mathrm{II}_{\mathbf{D}} = \left(D_{kl} D_{lk}\right)^{1/2} \tag{3.5}$$

m is the power-law index and $\mathrm{II}_{\mathbf{D}}$ represents the second invariant of the rate of deformation tensor $\mathbf{D}$.

It was recognized by Speziale [130] in 1984 that the axial vorticity is the cause of the secondary flows independently of the explicit constitutive structure of the fluid. The fluids under consideration can be put in context as all simple fluids in the sense of Truesdell and Noll [115]

$$\mathbf{T}(t) = -p(\rho)\mathbf{1} + \mathfrak{R}_{s=0}^{\infty}\{\mathbf{C}_t(t-s) - \mathbf{1}; \rho\}$$

where $\mathfrak{R}$ is an isotropic tensor valued functional and $\mathbf{C}_t(s)$, $\mathbf{1}$, p, and ρ are the relative Cauchy–Green tensor, the unit tensor, the pressure, and the density, respectively. Without assigning an explicit form to the extra stress, the curl of the linear momentum balance yields the defining equation for vorticity tensor ζ,

$$\rho\left(\zeta_{l,t} + u_m \zeta_{l,m}\right) = -\rho \varepsilon_{ijk} u_{k,m} u_{m,j} + \varepsilon_{lmn} S_{kn,mk}$$

the axial component of which reads

$$\rho\left[u_x \frac{\partial \zeta_z}{\partial x} + u_y \frac{\partial \zeta_z}{\partial y}\right] = \frac{\partial^2\left(S_{yy} - S_{xx}\right)}{\partial x \partial y} + \frac{\partial^2 S_{xy}}{\partial x^2} - \frac{\partial^2 S_{xy}}{\partial y^2},$$

where z is the axial coordinate. The secondary flow (u_x, u_y) is the direct result of the axial vorticity ζ_z if the axial vorticity source term is non-zero,

$$\frac{\partial^2 (S_{yy} - S_{xx})}{\partial x \partial y} + \frac{\partial^2 S_{xy}}{\partial x^2} - \frac{\partial^2 S_{xy}}{\partial y^2} \neq 0.$$

Thus, it is *necessary* that either $S_{yy} - S_{xx} \neq 0$ or $S_{xy} \neq 0$ be non-zero for the secondary flows to take place. That these are *necessary* rather than *sufficient* conditions was pointed out much later by Huang and Rajagopal [131].

In a recent paper, Yue et al. [132] extended Oldroyd's criteria for the existence of secondary flows to prove that secondary flows are not driven directly by the second normal stress difference N_2 but by the curl of an effective body force arising from N_2. They show that this body force has to be non-conservative for the secondary circulation to exist. The former requires that the second normal stress coefficient is not a constant multiple of the shear-dependent viscosity and the cross section is not axisymmetric. Further, they prove a general criterion in two parts for the existence and the direction of the secondary circulation, the first part of which is the same as the Oldroyd criterion (c) above, that is, if the inverse of the ratio m of the second normal stress coefficient $\Psi_2\left(\dot{\gamma}\right)$ and the shear-dependent viscosity $\eta\left(\dot{\gamma}\right)$ is an increasing function of the shear rate $\left(\dot{\gamma}\right)$

$$\frac{\mathrm{d}(m^{-1})}{\mathrm{d}\dot{\gamma}} = \frac{\mathrm{d}}{\mathrm{d}\dot{\gamma}}\left[\frac{\Psi_2\left(\dot{\gamma}\right)}{\eta\left(\dot{\gamma}\right)}\right] > 0 \tag{3.6}$$

the secondary circulation will proceed from the region of high shear to the region of low shear along the wall. However, if the opposite occurs, the secondary flow will take place from low-shear regions to high-shear regions. The point is made that even though in all known cases of polymeric fluid flows Eq. (3.6) is satisfied and flow takes place from high-shear regions to low-shear regions, there is a whole new host of complex fluids including flexible polymer solutions, surfactants, liquid crystalline polymers, emulsions, foams, and suspensions, some of which may violate criterion (Eq. 3.6) and the opposite may be satisfied flow taking place from low shear to high shear. Further, remarks on this issue on suspensions are in Sect. 6.2.5.

3.5 Secondary Flows of Dilute Solutions: Rotating Pipes and Channels

3.5.1 Spanwise Rotating Pipes and Drag Reduction in Laminar Flow

An interesting feature of dilute polymer solutions with small relaxation times of the order of 10^{-3} s is the interaction of the secondary flow structure due to elastic effects with those stemming from the spanwise rotation of the straight tube. It has been shown that even very low rotation rates like the diurnal spin rate of the earth (7.292×10^{-5} rad/s) which corresponds to an Ekman number of 0.737 and which imposes a weak spanwise rotation on the square and/or rectangular tube generate secondary flows with strength of comparable or greater order of magnitude than the strength of the secondary flows induced by the viscoelasticity of the fluid in the case of dilute polymeric solutions with Weissenberg numbers We in the range of $10^{-5} < We < 10^{-2}$, Speziale and Thangam [133] and Thangam and Speziale [134]. Thus, the presence of spanwise rotations, albeit quite weak, can give rise to a double-vortex structure secondary flow, akin to that observed in ducts with a mild streamwise curvature, imposing a severe distortional effect on the eight-cell vortex structure due to elastic effects in the cross section. This distortion may result in a complete overshadowing of the viscoelastic secondary flows and lead to a degeneration into a predominantly double-vortex configuration in the cross section. This interesting result indicates that it is rather difficult to experimentally reproduce the eight-vortex secondary flow structure in laboratory experiments with dilute polymeric solutions. In addition, it appears that neglecting the effect of the earth rotation in the calculation of the secondary flows of dilute polymeric solutions is a questionable assumption.

When a long rectangular or square straight tube is filled with a dilute polymeric solution flowing under the influence of an applied pressure gradient in fully established flow mode in a rotating frame, the relevant dimensionless parameters are the Reynolds Re, Ekman Ek, and Weissenberg We numbers,

$$Re = \frac{\overline{w}D}{\nu}, \qquad Ek = \frac{\nu}{2\Omega D^2}, \qquad We = \frac{\lambda\mu}{2\rho D^2}$$

where $\overline{w}$, D, ν, Ω, and λ represent the average longitudinal velocity, the channel width, the kinematic viscosity, the angular rotation, and the relaxation time of the fluid. In an inertial frame (zero spanwise rotation $\Omega = 0$ or infinite Ekman number $Ek = \infty$), the secondary flow in the cross plane due to elastic effects has a structure made up of eight vortices independently of the aspect ratio of the tube and for any Reynolds number as long as the laminar nature of the flow is preserved. In a square tube, the vortices are symmetric with respect to the symmetry axes of the square, but in a rectangular cross section, the symmetry with respect to the diagonals is lost. In an inertial framing, the strength of the secondary flows increases with increasing Weissenberg numbers in the range of interest for dilute solutions say $10^{-5} < We < 10^{-3}$. When a very weak spanwise rotation is imposed equal to the diurnal spin rate of the earth ($\Omega = 7.292 \times 10^{-5}$ rad/s, $Ek = 0.737$), the strength of the secondary flow is seen to increase further as compared to the strength in the inertial frame, but remains constant for increasing We in the range $10^{-5} < We < 10^{-3}$, and any Reynolds number in the range ($10 < Re < 10^3$) (Fig. 3.4). This result suggests that secondary flows due to the effect of the earth's rotation of comparable magnitude to those due to elastic effects and of opposite sign arise to keep the magnitude of the combined secondary field constant with increasing We independently of the driving pressure gradient, that is, the Reynolds

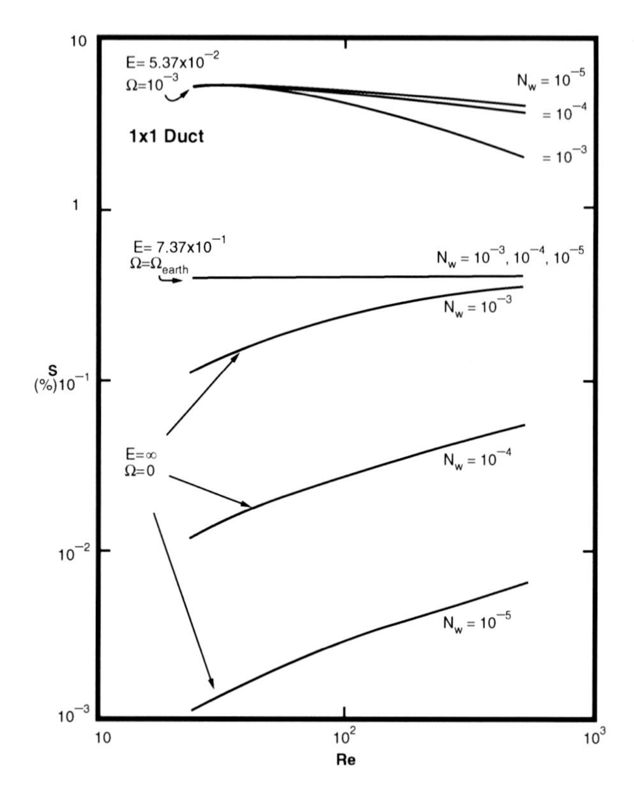

Fig. 3.4 Strength S of the secondary flow (as a % ratio of $[|u|,|v|]_{max}/|w|_{max}$) as a function of the Reynolds Re and Weissenberg $We = N_w$ (in the notation of [134] and thus in the same notation in Fig. 3.4) numbers in a square duct for different values of the rotation rate Ω (reprinted from Thangam and Speziale [134] with permission)

number. If the rotation rate is increased further, but is still in the moderate range ($\Omega = 10^{-3}$ rad/s, $Ek = 5.37 \times 10^{-2}$), a decrease in the strength of the combined secondary field is computed with increasing We in the range specified for the range of the Re considered. In fact, the larger the Re at a fixed We, the larger the magnitude of the decrease, and the larger the We at a fixed Re again the larger the decrease (Fig. 3.4). This is a very interesting phenomenon, the physical mechanism of which has been clarified due to the pioneering investigations of Speziale and Thangam [135]. In Fig. 3.4, the strength of the secondary flow is defined as the ratio ($[|u|, |v|]_{max}/|w|_{max}$) where u, v, and w represent the radial, azimuthal, and longitudinal components of the velocity.

By mid-1980s, there was an overwhelming accumulation of evidence that a weak spanwise rotation gives rise to a double-vortex secondary motion in the laminar duct flow of Newtonian fluids, Hart [136], Lezius and Johnston [137], Speziale [138], and Speziale and Thangam [139]. These studies were followed by the work of Speziale and Thangam [135] and Thangam and Speziale [134] who for the first time looked at the interaction of Coriolis and viscoelastic forces in dilute polymer solutions, neglecting shear-thinning in [135], which would hold for highly dilute viscoelastic fluids, and including shear-thinning in [134]. The non-linear stress response function of the fluid is represented in [134] by the corotational Maxwell model (Eq. 3.3) with the Jaumann derivative (Eq. 3.4) and the apparent viscosity obeying a truncated power law of the type (Eq. 3.5). They find that shear thinning has a surprisingly small effect of the flow structure. In either case, they find that the mere presence of the diurnal rotation of the earth gives rise to a double-vortex secondary flow due to Coriolis effects of a comparable or larger order of magnitude than the secondary flows caused by viscoelastic forces in the range of Weissenberg numbers $10^{-5} < We < 10^{-3}$ in the study, completely overshadowing and overwhelming the eight-cell structure in the cross plane resulting from visco-elastic forces in an inertial frame, the superposition degenerating into a predominantly double-vortex configuration.

Speziale and Thangam [133] and Speziale [138] show, using a non-linear stability analysis and an upper convected single mode Maxwell model with shear-thinning represented by a truncated power law as well as a Rivlin–Ericksen fluid of grade two and an explicit finite difference technique, that at moderate rotation rates there is a reduction in dissipation, that is, a reduction in drag with increasing We and Re numbers, thereby giving an example for the first time of the counterpart of Toms phenomenon [37] (see Sect. 2.2.3) in laminar flows. They conclude that their computations "tend to indicate that this drag reduction phenomenon in dilute viscoelastic channel flow subjected to a spanwise rotation arises as a result of the relaxation time of the polymer aligning itself with the secondary flow so that the net axial vorticity is reduced along with the dissipation". Thus, energy is stored rather than dissipated causing a reduction in the strength of the secondary flows.

3.5.2 Rotating Channel Flow

They provide further proof of this new phenomenon of drag reduction in laminar flow in a related study of rotating channel flow. It is well known that superimposed rigid body rotations can lead to instabilities of the Taylor type characterized by the development of Taylor vortices whose axes are perpendicular to the axis of rotation in the laminar flow of Newtonian fluids. It has been established for some time now that at weak rotation rates, there is a double-vortex secondary flow in the rotating channel; at intermediate rotation rates, there is instability in the form of longitudinal roll cells; and at rapid rotation rates, there is a restabilization of the flow to a double-vortex Taylor–Proudman regime. In the latter, the axial velocity profiles are independent of the coordinate along the axis of rotation in the interior of the channel. That the flow will restabilize to a two-dimensional structure is very much in line with the Taylor–Proudman theorem for inviscid fluids which predicts that the slow steady flow in a rotating framework is two dimensional. Thus, Taylor–Proudman theorem also holds in a very approximate sense for Newtonian fluids with small viscosity provided that the flow is sufficiently far removed from solid boundaries. The contributions of Speziale [138] and Speziale and Thangam [139] are at the very foundations of this body of knowledge with linear fluids. Extending their work to non-linear fluids, Speziale and Thangam [133] and Thangam and

Speziale [134] described for the first time the interaction of elasticity and Coriolis-driven secondary flows of purely elastic dilute fluids (with shear-thinning effects neglected) including computations in the range where the Coriolis forces are dominant, that is, at rotation rates of an order of magnitude greater than that of the diurnal spin rate of the earth. Through a non-linear stability analysis, they show, Speziale and Thangam [135] and Speziale [140], that the results obtained for linear fluids also hold qualitatively for dilute viscoelastic fluids. In particular for sufficiently large *Re* numbers, they show that instability in the form of longitudinal roll cells occurs at intermediate rotation rates. They demonstrate that the introduction of a very small amount of polymeric additive raising the *We* number to order 10^{-5} gives rise to secondary flows with a substantially reduced drag and has a stabilizing effect on the rotating channel flow. The dimensional equations which define the perturbation (u, v, w) to the base flow are

$$u\frac{\partial w}{\partial x} + v\frac{\partial w}{\partial y} = -\frac{1}{\rho}\frac{\partial P}{\partial z} + \nu\nabla^2 w + 2\Omega u$$

$$u\frac{\partial \zeta}{\partial x} + v\frac{\partial \zeta}{\partial y} = \nu\nabla^2\zeta + 2\lambda\mu\left(\frac{\partial w}{\partial x}\frac{\partial \nabla^2 w}{\partial y} - \frac{\partial w}{\partial y}\frac{\partial \nabla^2 w}{\partial x}\right) + 2\Omega\frac{\partial w}{\partial y}$$

$$\nabla^2\psi = \zeta, \qquad u = -\frac{\partial \psi}{\partial y}, \qquad v = \frac{\partial \psi}{\partial x},$$

where ψ and ζ represent the stream function in the cross section and the axial vorticity, respectively. The dimensionless numbers influencing the phenomena are the Reynolds *Re*, Rotation *Ro*, and Weissenberg *We* numbers,

$$Re = \frac{\overline{w}D}{\nu}, \qquad Ro = \frac{\Omega D}{\overline{w}}, \qquad We = \frac{\lambda\nu}{D^2}$$

where $\overline{w}$, D, ν, Ω, and λ represent the average longitudinal velocity, the channel width, the kinematic viscosity, the angular rotation, and the relaxation time of the fluid. Figure 3.5 shows the cell configurations including roll cells in the cross section of the channel for increasing *We* numbers. All roll cells in the interior of the channel disappear at a critical $We = 7.5 \times 10^{-5}$ for the *Re* and *Ro* numbers in Fig. 3.5. As the *Re* and the *Ro* numbers are increased keeping the viscosity constant and the flow in the laminar regime, the value of the critical *We* decreases.

To quantify the friction losses resulting from the secondary flows, they use a flow rate reduction parameter $(Q - Q_r)/Q$ where Q and Q_r are, respectively, the flow rate in a stationary channel without secondary flows and the flow rate in a rotating channel. Figure 3.6 represents the variation of this parameter with *We*. An increase in the *We* leads to a substantial decrease in the flow rate reduction, and most of that reduction occurs in the unstable regime with longitudinal roll cells.

The introduction of a minute amount of a long chain polymer into a Newtonian fluid has a stabilizing effect on rotating channel flow and gives rise to secondary

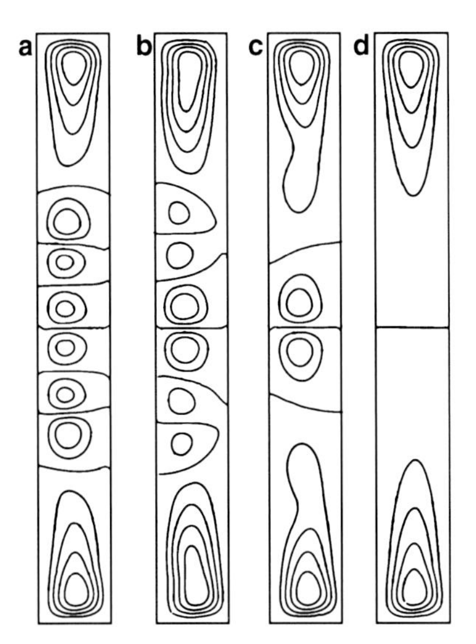

Fig. 3.5 Fully developed secondary flow streamlines in an 8 × 1 channel (reprinted from Speziale and Thangam [135] with permission): $G = -P_{,z} = 0.2 \times 10^{-3}$ lb/ft^3, $\Omega = 5 \times 10^{-3}$. (**a**) $Re = 248$, $Ro = 0.047$, $We = 0$. (**b**) $Re = 252$, $Ro = 0.046$, $We = 2.25 \times 10^{-5}$. (**c**) $Re = 256$, $Ro = 0.046$, $We = 5.59 \times 10^{-5}$. (**d**) $Re = 248$, $Ro = 0.047$, $We = 9.02 \times 10^{-5}$

Fig. 3.6 Variation of the flow rate reduction parameter $\frac{Q-Q_r}{Q}$ in a rotating channel as a function of the Weissenberg number We (reprinted from Speziale and Thangam [135] with permission). $G = -P_{,z} = 0.2 \times 10^{-3}$ lb/ft^3, $\Omega = 5 \times 10^{-3}$

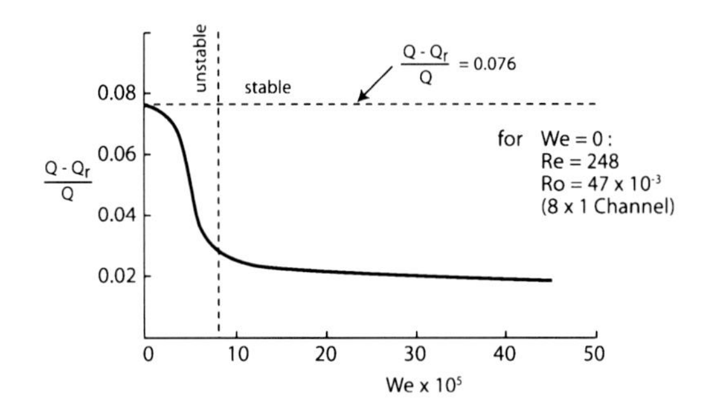

flows with a substantially reduced frictional drag approximately one-third of that of a Newtonian fluid. The overwhelming majority of this drag reduction occurs in the unstable regime. Similar effects have been observed in the Couette flow of dilute polymer solutions by Denn and Roisman [141], Denn et al. [142] and Ginn and Denn [143]. That this drag reduction occurs as a result of the relaxation time of the polymeric additive aligning itself with the secondary flow so that energy is stored rather than dissipated and that this drag reduction similar to Toms phenomenon is more pronounced in small channels rather than in large channels similar to drag reduction in small and large diameter tubes (see Sect. 2.2.3.3 as well as Berman [80]) are shown by examining the axial vorticity transport equation which reads in dimensionless form

$$u\frac{\partial \zeta}{\partial x} + v\frac{\partial \zeta}{\partial y} = \frac{1}{Re}\nabla^2\zeta + 2We\left(\frac{\partial w}{\partial x}\frac{\partial \nabla^2 w}{\partial y} - \frac{\partial w}{\partial y}\frac{\partial \nabla^2 w}{\partial x}\right) + 2Ro\frac{\partial w}{\partial y}.$$

They show computationally that the viscoelastic axial vorticity source term

$$2We\left(\frac{\partial w}{\partial x}\frac{\partial \nabla^2 w}{\partial y} - \frac{\partial w}{\partial y}\frac{\partial \nabla^2 w}{\partial x}\right)$$

leads to a reduction in the average value of the axial vorticity squared ζ^2 and hence a reduction in the dissipation of the secondary flow when We is different than zero.

These milestone investigations give a clear answer to some of the fundamental questions posed in the comprehensive review article by Hartnett and Kostic [144]. They ask if very dilute concentrations of drag-reducing aqueous polymer solutions are viscoelastic Newtonian fluids [144, p. 345]. The question arises because they point out that the energy loss in the laminar longitudinal flow of these fluids can be predicted using Newtonian assumptions, that is, there is no drag-reducing effect in laminar flow, but there is a strong drag-reducing effect in turbulent flow, that is, pressure drop does not agree with Newtonian predictions. They raise the additional question as to whether such dilute polymer solutions generate secondary flows in the laminar flow through non-circular tubes. Speziale and Thangam show that indeed there is a drag-reducing effect in the laminar flow of dilute solutions and there are secondary flows.

3.6 Recent Investigations of Secondary Flows of Viscoelastic Fluids

3.6.1 *Steady Flows in Straight Tubes*

There are no studies of secondary flows from the mid-1970s to the early 1990s except those of Speziale and Thangam [135] and Thangam and Speziale [134] and Speziale [140] summarized in the previous paragraphs. Significant progress was made in developing numerical methods appropriate to tackle constitutively non-linear fluid dynamics problems in the 1980s, Crochet et al. [145] and Keunings [146]. The effort to develop robust numerical algorithms to solve in particular high Weissenberg or Deborah number flows is a continuing focal point of research. These developments stimulated research and several papers investigating secondary flows appeared in the archival literature starting from the early 1990s. The overwhelming majority of these investigations are numerical. A decade and a half after the efforts of Dodson et al. [127] and Townsend et al. [128] in the first comprehensive study of secondary flows in straight ducts, Gervang and Larsen [147] use the same CEF equation with variable coefficients, Criminale et al. [114], used by Dodson et al. [127] to investigate the secondary flows of a real fluid, a 2 %

viscarin solution in distilled water for which rheometric functions η, Ψ_1, Ψ_2 were available and taken as power-law point functions of the deformation field,

$$\eta = k\left(2tr\mathbf{D}^2\right)^{(n-1)/2}, \quad k = 8.5Pa\, s^n, \quad n = 0.37,$$
$$\Psi_1 = k\left(2tr\mathbf{D}^2\right)^{n_1/2}, \quad k = 5.96Pa\, s^{n_1+2}, \quad n_1 = -1.35,$$
$$\Psi_2 = -0.15\Psi_1,$$

in rectangular tubes and offer theoretical predictions based on a finite volume solution of the full field equations and experimental verification via LDA measurements of the velocity components in the cross section. The numerical solution employs a three-dimensional discretization on strained rectangular grids, the PISO (Pressure Implicit with Splitting of Operators) algorithm for the velocity–pressure coupling, an ADI solver for the momentum equations, and an incomplete Cholesky factorization in the preconditioned conjugate gradient solver for the pressure equations. A remark about the particular CEF constitutive equation used in this paper is in order. The rheometric functions η, Ψ_1, Ψ_2 were defined as point functions in the field, which guarantees that the CEF constitutive equation will predict secondary flows; if the rheometric functions were taken as constants, the equation would not have predicted secondary flows (see Sect. 3.4). Whether or not the predictions will match experimental observations is an entirely different matter. They predict a structure with 8 vortices, 2 counter-rotating in each quadrant, for small and moderate aspect ratios and a pattern with 12 vortices, 3 in each quadrant, 1 albeit very weak, for large aspect ratios. Specifically they investigate the secondary field in four rectangular cross sections varying from aspect ratio 1 to 16 the width being equal to or greater than the height, all with the same cross-sectional area of 16 mm^2. They determine that fluid particles move along sidewalls to leave from corners, which explains the physical mechanism behind the secondary vortices, along the lines first observed experimentally by Giesekus [107] in 1965 in elliptical cross sections. They also determine that the conclusions reached by Townsend et al. [128] hold.

The mean longitudinal velocity in the duct varies from 0.1 m/s to as high as 10 m/s. In an important experimental verification, they show that longitudinal velocity profiles are well predicted by power-law models when the mean velocity in the rectangular duct is below 3 m/s and that the energy losses due to secondary flows are less than 2–4 % of the total pressure drop. Hence, for this range of shear rates, secondary flows have a negligible effect on the energy loss. However, they also show that for higher shear rates when the mean velocity varies from 5 to 10 m/s, there appears a significant increase in pressure drop secondary flows contributing from 6 to 20 % of the total pressure drop; in other words, the pressure gradient shows an increase 6 to 20 % over that predicted by the power law.

Xue et al. [148] use the MPTT (modified Phan-Thien–Tanner) model, a more realistic constitutive structure proven to be reasonably successful in predicting at least qualitatively a variety of complex flows to take a detailed look at the structure of secondary flows in rectangular tubes of various aspect ratios via a new variant of the finite volume method introduced for this purpose with the emphasis on the numerical algorithm. The MPTT model is part of a family of non-affine constitutive formulations framed in terms of the non-affine Gordon–Schowalter convected derivative $\overset{\circ}{\mathbf{S}}$

$$\mathbf{T} = -p\mathbf{1} + 2\eta_N \mathbf{D} + \mathbf{S} \tag{3.7a}$$

$$f[tr\mathbf{S}]\mathbf{S} + \lambda \overset{\circ}{\mathbf{S}} = 2\eta_m \mathbf{D} \tag{3.7b}$$

$$\overset{\circ}{\mathbf{S}} = \overset{\bullet}{\mathbf{S}} - \left(\nabla \mathbf{u}^{\mathrm{T}} - \xi \mathbf{D}\right)\mathbf{S} - \mathbf{S}\left(\nabla \mathbf{u}^{\mathrm{T}} - \xi \mathbf{D}\right)^{\mathrm{T}} \tag{3.7c}$$

$$\overset{\circ}{\mathbf{S}} = \overset{\bullet}{\mathbf{S}} - \boldsymbol{\zeta}^{\mathrm{T}}\mathbf{S} - \mathbf{S}\boldsymbol{\zeta} - \alpha(\mathbf{DS} + \mathbf{SD}), \quad \alpha = 1 - \xi \neq -1, 0, 1$$

$$f(\varepsilon_o, tr\mathbf{S}) = \exp\left(\frac{\varepsilon_o \lambda}{\eta_{mo}} tr\mathbf{S}\right)$$

$$\eta_m = \eta_{mo}\left[1 + \xi(2-\xi)\lambda^2\left(\overset{\bullet}{\gamma}\right)^2\right]\left[1 + \lambda^2\left(\overset{\bullet}{\gamma}\right)^2\right]^{(n-1)/2} \qquad \overset{\bullet}{\gamma} = \left(2tr\mathbf{D}^2\right)^{1/2}$$

η_m, η_N, η_{mo}, ε_o, ξ, n, λ, $\overset{\bullet}{\gamma}$, $\overset{\bullet}{\mathbf{S}}$, $\mathbf{u}$, and $\boldsymbol{\zeta}$ stand, respectively, for the shear rate-dependent molecular viscosity, the Newtonian viscosity of the solvent, the zero shear rate molecular viscosity, the constitutive constant which directly determines extensional properties, the constitutive constant which defines the degree of slippage experienced by long chains vis-à-vis the surrounding Newtonian continuum, the power index, the relaxation time of the fluid (essentially of the long chains), the shear rate, the material derivative of the extra-stress tensor, the velocity field, and the vorticity tensor with $\mathbf{T}$, $\mathbf{S}$, $\mathbf{D}$, and p representing the total stress, the extra-stress tensor, the rate of deformation tensor, and the constitutively indeterminate part of the total stress pressure. ε_o is the constitutive parameter which defines the elongational properties as well as the shear-thinning ability. It should be noted that both ε_o and ξ govern shear-thinning. If $\varepsilon_o = 0$, that is, $f = 1$, we obtain a model which can describe the response of fluids to forcing which lead to negligible elongational deformations. In the absence of ε_o, the constitutive parameter ξ governs the shear thinning as well as the slippage. The relative magnitudes of ε_o and ξ indicate the extent to which each controls shear-thinning. For instance, if $\varepsilon_o = 0$, shear-thinning is entirely governed by ξ, that is, by the slippage of the polymer strands with respect to the surrounding continuum. The deformation of the macroscopic medium is affine by definition, but polymer strands embedded in the medium may slip with respect to the deformation of the macroscopic medium. Thus, each strand may transmit only a fraction of its tension to the surrounding continuum as the continuum slips past the strands. This kind of slippage is taken into account in the

Gordon–Schowalter convected derivative by the material parameter ξ. When $\xi = 0$, there is no slippage, the motion becomes affine, the Gordon–Schowalter derivative collapses onto the upper convected derivative, and shear-thinning is entirely governed by the material parameter ε_o. The power-law index n is always $n \leq 1$. The zero shear rate viscosity of the fluid is defined as $\eta_o = \eta_{No} + \eta_{mo}$. η_{No} is the zero shear rate viscosity of the Newtonian solvent. The constitutive equation (Eq. 3.7a) can then be rewritten as

$$\mathbf{T} = -p\mathbf{1} + 2\eta_o(1-\beta)\mathbf{D} + \mathbf{S}, \quad \beta = \frac{\eta_{mo}}{\eta_o}$$

In a simple shearing flow with the indices (1) and (2) representing the direction of the shear and the direction perpendicular to the shear planes, respectively, the viscometric functions, that is, the shear stress $\tau\left(\dot{\gamma}\right)$ and the first and the second normal stress differences $N_1\left(\dot{\gamma}\right)$ and $N_2\left(\dot{\gamma}\right)$ implied by the constitutive structure (Eq. 3.7a, b, c), can be deduced from the viscosity function $\eta\left(\dot{\gamma}\right)$ and the first and second normal stress coefficients $\Psi_1\left(\dot{\gamma}\right)$ and $\Psi_2\left(\dot{\gamma}\right)$,

$$\eta\left(\dot{\gamma}\right) = \eta_N\left(\dot{\gamma}\right) + \frac{f(\varepsilon_o, tr\mathbf{S})\eta_m\left(\dot{\gamma}\right)}{f^2(\varepsilon_o, tr\mathbf{S}) + \xi(2-\xi)\lambda^2\dot{\gamma}}$$

$$\Psi_1\left(\dot{\gamma}\right) = \frac{N_1}{\dot{\gamma}} = \frac{S_{11} - S_{22}}{\dot{\gamma}} = \frac{2\lambda}{f(\varepsilon_o, tr\mathbf{S})}\left[\eta\left(\dot{\gamma}\right) - \eta_N\left(\dot{\gamma}\right)\right]$$

$$\Psi_2\left(\dot{\gamma}\right) = \frac{N_2}{\dot{\gamma}} = \frac{S_{22} - S_{33}}{\dot{\gamma}} = -\frac{\xi}{2}\Psi_1\left(\dot{\gamma}\right).$$

They use a stable and efficient implicit finite volume method (FVM) constructed on a variant of the SIMPLER (Semi-Implicit Method for Pressure-Linked Equations Revised) algorithm invented by Patankar [149]. The modified SIMPLER method they develop is christened SIMPLEST (SIMPLE with a new splitting technique). The SIMPLEST algorithm avoids some of the shortcomings of the original SIMPLE as well as some of its later variants such as SIMPLEC, Van Doormaal and Raithby [150]. A persistent problem with SIMPLE and SIMPLEC is for time-dependent flows very small time steps must be chosen as the pressure and velocity fields obtained at the end of each time step do not satisfy one and the same momentum balance, or alternatively iteration is a necessity at each time step; otherwise, it is questionable whether the fields are legitimate approximations to the solution of the FV difference equations over the time step, Issa [151].

They predict a structure with 8 vortices in the cross section for all aspect ratios which contradicts the 12 vortex structure predictions of Gervang and Larsen [147] for large aspect ratios (Fig. 3.7).

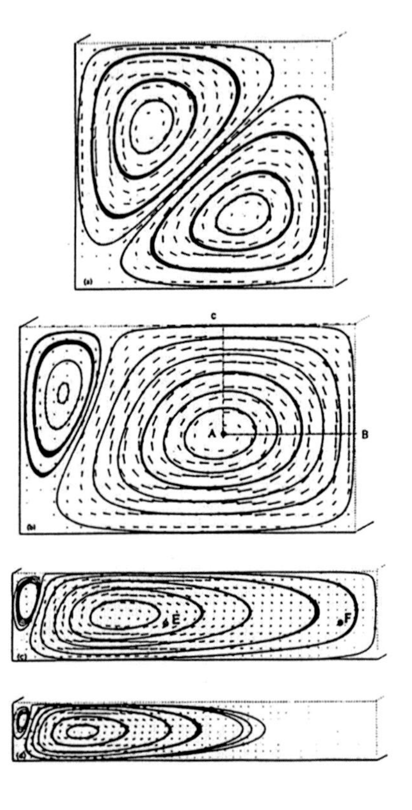

Fig. 3.7 Streamlines of secondary flow in one quarter of the cross section of a rectangular pipe with aspect ratio of 1, 1.56, 4, and 6.25 (reprinted from Xue et al. [148] with permission)

An increase in the value of the constitutive parameter ε_o increases the strength of the secondary flows. This is due to the shear-thinning behavior of the fluid becoming more marked which increases the strength of the primary longitudinal flow as well as increasing the departure of Ψ_2 from a fixed constant multiple of η. In the same way, the material parameter ξ affects the strength of the secondary flows as it directly determines the strength of Ψ_2 for the MPTT fluid and through the strength of the primary flow as it also controls the viscosity (shear-thinning). The extent to which it controls the viscosity is determined by the relative values assigned to ξ and ε_o. Thus, an increase in the value assigned to the slip parameter ξ also corresponds to an increase in the strength of the secondary flow. For a fixed level of elasticity and given pressure gradient increasing the viscosity slows down the primary flow and weakens the secondary flow. This can be accomplished by increasing the zero shear rate viscosity of the power exponent n.

Multimode differential models are increasingly considered as good candidates for the simulation of complex industrial flows. Debbaut et al. [152] and Debbaut and Dooley [153] reported comparisons between experiments with low-density polyethylene (LDPE) and finite element predictions for an N-mode Giesekus model. The former authors study secondary flows in square, rectangular, and teardrop-shaped channels, whereas the latter investigate the flow in rectangular and tapered channels with square cross sections by solving numerically

$$\nabla \bullet \mathbf{T} = \nabla p, \qquad \nabla \bullet \mathbf{u} = 0 \tag{3.8}$$

$$\mathbf{T} = \sum_{i=1}^{N} \mathbf{T}_i$$

$$\mathbf{T}_i \left[\mathbf{1} + \frac{\alpha_i \lambda_i}{\eta_i} \mathbf{T}_i \right] + \lambda_i \mathbf{T}_i = 2\eta_i \mathbf{D} \tag{3.9}$$

where a discrete spectrum of N relaxation modes are considered with λ_i, η_i, and α_i representing, respectively, the relaxation times and the partial viscosity factors of the spectrum and the ratio of the second to the first normal stress differences or coefficients in the spectrum. $\mathbf{T}_i$ is the contribution of the ith relaxation mode to total viscoelastic stress $\mathbf{T}$. Similar spectra representing different batches of LDPE are used in both papers with four and five relaxation modes, respectively, in [152] and [153]. The spectra are selected in view of the available experimental data. For instance, in Debbaut and Dooley's work, the parameters assume the following values: λ_i (s) = 1, 0.1, 0.01, 0.001; η_i (poise) = 33,711, 17,345, 5,467, 1,752; and α_i = 0.5, 0.6, 0.7, 0.8. They use the elastic viscous split stress (EVSS) algorithm developed by Rajagopalan et al. [148] which splits each individual contribution $\mathbf{T}_i$ to the total viscoelastic extra-stress tensor $\mathbf{T}$ (total stress minus modified pressure p) into elastic and viscous contributions $\mathbf{T}_i = 2\eta_i\mathbf{D} + \mathbf{S}_i$. Then Eq. (3.9) can be rewritten for individual $\mathbf{S}_i$, and the full momentum equation (Eq. 3.8)$_1$ is expressed in terms of a purely diffusive component with positive consequences in the computations, Rajagopalan et al. [154].

$$\nabla \bullet \left[\sum_{i=1}^{N} \mathbf{S}_i \right] + 2 \left[\sum_{i=1}^{N} \eta_i \right] \nabla \bullet \mathbf{D} = \nabla p$$

In their numerical algorithm, they use slip at the wall as a numerical tool for solving the set of field equations through a continuation procedure by progressively decreasing the slip coefficient β_s at the wall from a relatively large value down to no-slip, that is, to zero, and present only the final solutions for the no-slip condition at the wall.

Debbaut et al. [152] determine that in the single moderately low aspect ratio (height smaller than width) rectangular channel they investigated, there are three stagnation points and eight symmetrically located cells with four dominant cells each about 35 times stronger than four small and weak cells squeezed along the sidewalls. Six cells are found in the teardrop-shaped channel with two stagnation points where four cells merge. The experimental setup used in both papers is identical and consists of two screw extruders attached directly to a co-extrusion feedblock which in turn is attached to a die channel (Fig. 3.8).

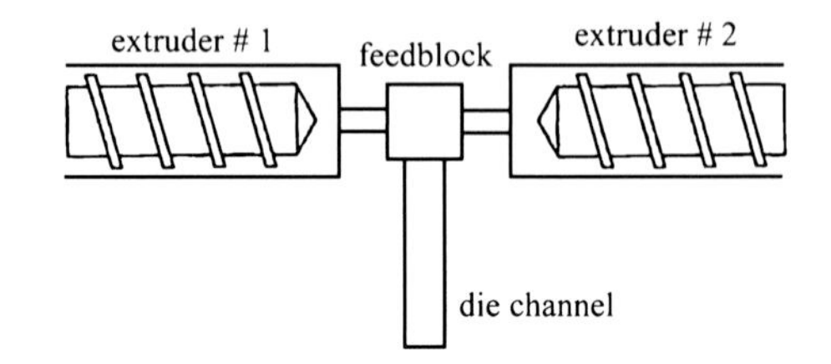

Fig. 3.8 Co-extrusion feedblock attached to a die channel

Two batches of the same LDPE melt, one with white and the other with black pigmentation, are used. The pigmentation does not affect the properties of the melt significantly if at all. The black-and-white-pigmented LDPE is fed into the feedblock in a well-controlled flow rate ratio of 4 to 1, respectively, at a temperature of 204 °C. They note that the respective areas occupied by the black and white fluids differ from the flow rate ratio of 4:1 because the mean axial velocity of the black fluid is higher than that of the white layer. The interface between layers in the die channel is tracked computationally using a method developed by Avalosse [155] by tracking the path of a large number of arbitrarily selected fluid particles in [152], whereas in [153], a pure advection equation is used for tracking the fluid motion. Experimentally after steady-state conditions are reached, extruders are idled and the flow is stopped. When the material in the die channel is cooled and becomes solid, the square rod of solid material is removed from the channel and is cut at specified intervals along the longitudinal flow direction. Experimental sections are slightly different for the investigations reported in [152] and [153], but the die channel length is the same 61 cm. Comparisons of the experimentally determined interface deformations with theoretically computed interface shapes are offered for the square, rectangular, and teardrop shapes in [152] for one flow rate and the square straight channel for a low and a high value of the flow rate in [153] and tapered square channels as well for one flow rate. Qualitative predictions are good and follow the pattern observed in the experiments, but the theoretical model underestimates the deformation of the interface in all cases. The higher the shear rate, the more qualitative the predictions become. For instance, for the higher flow rate in the square straight channel, the deformation predicted is much smaller than that observed in all experimental cross sections along the die; same is true for the square-tapered channel as well. This can be attributed to the dissipative unstable Giesekus equation at the very basis of the computations (see Siginer [73] for the dissipative instability of the Giesekus constitutive equation). The growing influence of the dissipative terms in Eq. (3.9) at higher flow rates ultimately leading to instability is distorting the predictions and widening the gap with experimental observations. By contrast, secondary flows have a much more pronounced effect in the tear-drop-shaped channel (Fig. 3.9).

An interesting numerical experiment conducted in [153] clearly demonstrates the dramatic effect of the secondary vortices on the overall flow and how important a consequence that may carry for complex industrial flows. The interfaces in a multilayered fluid system conceived of as alternating black-and-white-pigmented layers horizontally co-extruded in the beginning of the die are studied by tracking

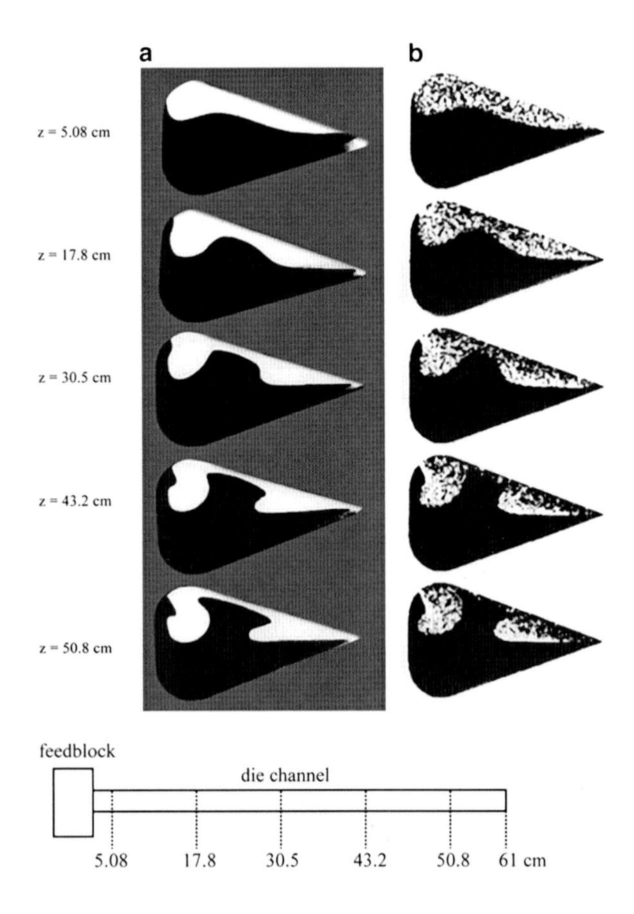

Fig. 3.9 Flow through the teardrop channel (reprinted from Debbaut et al. [152] with permission): (**a**) Experimental results at the indicated locations in the die channel. (**b**) Numerical predictions at the indicated locations in the die channel

an advected variable. The initial horizontal structure of the layers is gradually lost, and at the end of the die, the interface structure is drastically different (Fig. 3.10).

In another attempt to bring clarity to the impact of secondary flows on industrial processes, Siline and Leonov [156] adopt a completely different analytical approach and decompose the general flow problem into a quasi-rectilinear component (the longitudinal component) and disturbances which create the secondary flows. They set out to show that for many applications involving polymer processing, secondary flows can be ignored or can be treated as small disturbances relative to the main almost rectilinear axial flow component given that the velocity components in the cross section are typically two orders of magnitude smaller than the primary axial velocity. The flow partitioning they propose is not restricted to any particular geometry and to small driving forces (small pressure gradients). The approach separates the driving forces into two parts: the pressure gradient which drives the primary flow and viscoelastic forces which drive the secondary flows. The primary flow is shaped by the non-Newtonian viscosity characteristics alone.

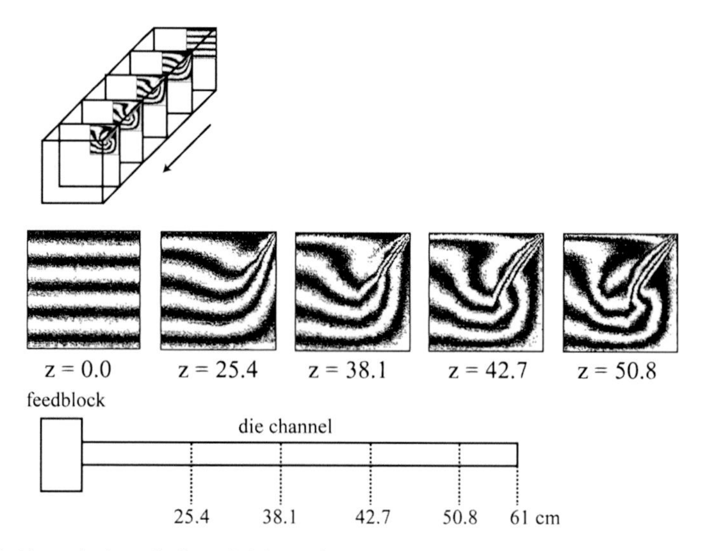

Fig. 3.10 Numerical prediction of deformations undergone by a pigmented, stratified multilayered system (material in the layers has the same density) at a flow rate of 2.5 kg/h (reprinted from Debbaut and Dooley [153] with permission)

They show that the viscoelastic force driving the secondary flows is small as compared to the primary driving force in both the slow and fast flow limits. The non-linear multimode viscoelastic constitutive model of the differential type they use has been proposed and tested by Leonov [157] to be Hadamard and dissipative stable (see Siginer [73] for the dissipative and Hadamard stability of the Leonov constitutive equation).

$$\mathbf{S} = \sum_{k=1}^{N} \mathbf{S}_k, \quad \mathbf{S}_k = G_k \left(\frac{tr\mathbf{C}_k}{3}\right)^n \mathbf{C}_k$$

$$2\lambda_k \overset{\nabla}{\mathbf{C}}_k + a\{\mathbf{C}_k^2 + b\mathbf{C}_k - \mathbf{1}\} = 0, \quad a = \exp\left\{m\left[\frac{tr\mathbf{C}_k}{3} - 1\right]\right\}, \quad b = \frac{tr\mathbf{C}_k^{-1} - tr\mathbf{C}_k}{3} \tag{3.10}$$

Here, $\mathbf{C}_k$ is the Finger elastic strain tensor in kth relaxation mode whose evolution is described by Eq. (3.10). G_k and λ_k are the linear Hookean elastic modulus and the relaxation time of the kth mode, and m and n are mode independent parameters. Simhambhatla and Leonov [158] in testing the predictions of (Eq. 3.10) use four relaxation modes to test five different polymer melts which include a high-density (HDPE) and low-density polyethylene (LDPE). They take $m = 1.4$ and $n = 0$ to adequately describe data in steady and unsteady shearing and extensional flows over three decades of deformation rates in simple shear, simple elongation, planar extension, as well as in biaxial extension. Tests in each case include start-up flow and stress relaxation except the simple shear which also includes creep and recovery. In particular, it was found that if the parameter m was assigned the value 1.4 extension stress growth data is adequately described. Data was well predicted except in the planar elongation case. They predict strain softening behavior in planar elongation, whereas strain hardening is observed experimentally. It may be appropriate here to define strain hardening and softening for the sake of completeness. The large increases in the elongational viscosity η_e over relatively small intervals of time shown in Fig. 3.11 are referred to as strain hardening, Meißner [159]. The opposite would be strain softening. Similar phenomena are not observed in steady shear. These large increases in the elongational viscosity under steady straining do not continue indefinitely but tend to a plateau. The beginning of this tendency at long times is visible in Fig. 3.11. where $\left(\dot{\varepsilon}\right)$ and $\left(\dot{\gamma}\right)$ represent the strain rate and shear rate, respectively.

This shortcoming was partially corrected in [157] where the following expression to model the dissipative term a in Eq. (3.10) is introduced:

$$a = \exp[-\beta(tr\mathbf{C} - 3)] + \frac{\sinh[\nu(tr\mathbf{C} - 3)]}{\nu(tr\mathbf{C} - 3)} - 1, \quad 0 \le \beta \le 1, \quad \nu \ge 0.$$

They recommend the use of $n = 0.1$ for very high values of the Deborah number higher than those in their experiments. $m = 1.4$ and $n = 0.1$ are adopted in the work of Siline and Leonov [156] on secondary flows. They expand the velocity field and the extra-stress tensor in asymptotic series,

$$\mathbf{u} = \mathbf{u}^{(0)} + \mathbf{u}^{(1)} + \mathbf{u}^{(2)} + O\left(\varepsilon^3\right),$$
$$\mathbf{S} = \mathbf{S}^{(0)} + \mathbf{S}^{(1)} + \mathbf{S}^{(2)} + O\left(\varepsilon^3\right)$$

and assume that $\mathbf{u}^{(0)} = [w(x_2, x_3) + w'(x_2, x_3)]\, \boldsymbol{e}_3$ where w and w' represent the full longitudinal velocity and the longitudinal disturbance, respectively, x_3 is the axial coordinate, and $\boldsymbol{e}_3$ is the physical unit vector in the axial direction. They propose the following partitioning of the flow

$$\mathbf{S}^{(0)}\left[\Re\left(w + w'\right)\right] = \mathbf{S}^{(0)}[\Re(w)] + \delta\mathbf{S}^{(0)}\left[\Re(w), \Re\left(w'\right)\right],$$
$$\Re(\bullet) = \mathbf{e}_2(\bullet)_{,2} + \mathbf{e}_3(\bullet)_{,3}$$

and assume that $|w'| \ll |w|, \quad |\delta\mathbf{S}^{(0)}| \ll |\mathbf{S}^{(0)}|$. Longitudinal velocity U is defined by a second-order non-linear elliptic boundary value problem with a unique solution

$$S^{(0)}_{12,2} + S^{(0)}_{13,3} = \Re \bullet \left\{\eta\left(\dot{\gamma}\right)\Re(w)\right\} = -p_{,3}, \quad w|_{\partial D} = 0,$$

where $\eta\left(\dot{\gamma}\right)$ is the shear rate-dependent viscosity with the generalized shear rate defined as $\dot{\gamma} = \sqrt{u^2 + v^2}$ and ∂D represents the contour of the non-circular cross

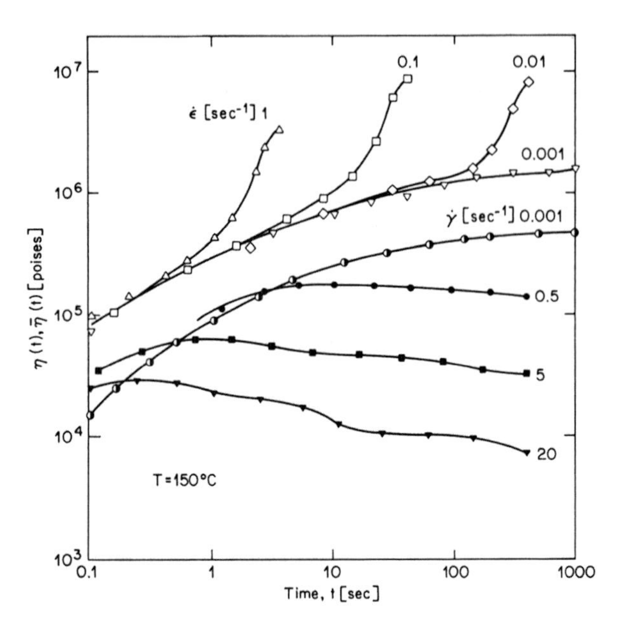

Fig. 3.11 Elongational viscosity $\eta_e = \bar{\eta}$ and shear viscosity $\eta_s = \eta$ as functions of time after inception of steady straining. Open symbols represent elongational viscosities; closed and half-closed symbols represent shear viscosities (reprinted from Meißner [159] with permission)

section. The first-order disturbances $\mathbf{u}^{(1)}$ and $\mathbf{S}^{(1)}$ are written in terms of the stream function ψ for the secondary flow

$$\mathbf{u}^{(1)} = \left\{ w', \psi_{,3}, -\psi_{,2} \right\}$$

$$\mathbf{S}^{(1)} = \delta\mathbf{S}^{(0)} \left\{ \mathfrak{R}(w), \mathfrak{R}\left(w'\right) \right\} + \mathbf{S}' \left\{ \psi_{,i}, \psi_{,ij}, \ldots \right\}$$

and are determined by

$$\mathfrak{R}\left(\mathbf{S}' + \delta\mathbf{S}^{(0)}\right) = 0,$$

$$\left(S'_{22} - S'_{33}\right)_{,23} - S'_{23,22} + S'_{23,33} = -\left\{ \left(S^{(0)}_{22} - S^{(0)}_{33}\right)_{,23} - S^{(0)}_{23,22} + S^{(0)}_{23,33} \right\}$$

The predictions for a square cross section are compared to those of Debbaut et al. [152] and claim quantitatively similar results. It is rather interesting that Debbaut et al. [152] who used in their calculations the dissipative unstable but Hadamard stable Giesekus equation and Siline and Leonov [156] who used an equation both Hadamard and dissipative stable (see Siginer [73] for Hadamard and dissipative instabilities) arrive at almost the same predictions. The reason of course lies in the Hadamard stability and the relatively unimportant contribution of the dissipative terms to the overall flow in the case of Debbaut et al. [152].

The fully developed steady velocity field in pressure-gradient-driven laminar flow of non-linear viscoelastic fluids with instantaneous elasticity constitutively represented by a class of single mode, non-affine quasilinear constitutive equations that include the MPTT and Johnson–Segalman models in straight pipes of arbitrary contour ∂D is investigated by Siginer and Letelier [15]. The constitutive structure connecting deformation measures represented by the rate of deformation tensor $\mathbf{D}$ with the viscoelastic contributed stress tensor $\mathbf{S}$ through a relaxation time λ, a molecular contributed viscosity η_m, and a function f related to the elongational properties of the fluid is considered.

$$\mathbf{T} = -p\mathbf{1} + 2\eta_N \mathbf{D} + \mathbf{S}$$

$$f(\varepsilon_o, tr\mathbf{S})\mathbf{S} + \lambda \overset{o}{\mathbf{S}} = 2\eta_m \mathbf{D}$$

$\mathbf{T}$, p, $\mathbf{S}$, and $\mathbf{1}$ represent the total stress, the constitutively indeterminate and determinate parts of the Cauchy stress tensor, and the unit tensor, respectively, and η_N refers to the Newtonian viscosity of the continuum. This structure representing a class of non-affine, quasilinear constitutive models which includes the Johnson–Segalman and the MPTT models is framed in terms of the Gordon–Schowalter convected derivative (interpolated Maxwell convected derivative) $\left(\overset{o}{\bullet}\right)$ (see Eq. 3.7c as well) with $\left(\dot{\bullet}\right)$ indicating the material derivative,

$$\begin{aligned} \overset{o}{\mathbf{S}} &= \dot{\mathbf{S}} - \nabla\mathbf{u}^{\mathrm{T}}\mathbf{S} - \mathbf{S}\nabla\mathbf{u} + \xi(\mathbf{SD} + \mathbf{DS}) \\ &= \mathbf{S}_{,t} + \mathbf{u} \bullet \nabla\mathbf{S} - \left(\nabla\mathbf{u}^{\mathrm{T}} - \xi\mathbf{D}\right)\mathbf{S} - \mathbf{S}\left(\nabla\mathbf{u}^{\mathrm{T}} - \xi\mathbf{D}\right)^{\mathrm{T}} \end{aligned}$$

If $\xi = 0$, the motion becomes affine, and Gordon–Schowalter convected derivative collapses onto the upper convected derivative. In that case, this structure will not predict secondary flows as the second normal stress difference $N_2\left(\dot{\gamma}\right) = 0$. If $\xi = 1$, the motion is again affine with the Gordon–Schowalter derivative merging into the corotational Jaumann derivative $\overset{o}{\mathbf{S}} = \dot{\mathbf{S}} - \boldsymbol{\zeta}^{\mathrm{T}}\mathbf{S} - \mathbf{S}\boldsymbol{\zeta}$ (see Eq. 3.4 for a different expression for the corotational Jaumann derivative) with $\boldsymbol{\zeta}^{\mathrm{T}}$ and $\boldsymbol{\zeta}$ referring to the transpose of the vorticity tensor and the vorticity tensor itself, respectively. This family of constitutive models has two constitutive parameters ε_o and ξ apart from the relaxation time λ and other constants embedded in the molecular viscosity η_m. The elongational and the non-affine deformation of the material are governed by ε_o and ξ, respectively. Both parameters also influence the shear-thinning behavior. In the latter instance, the relative magnitudes of ε_o and ξ indicate the extent to which each controls shear-thinning.

Siginer and Letelier [15] use a continuous one-to-one mapping to obtain arbitrary tube contours from a base tube contour ∂D_o. The one-to-one and continuous mapping determines the shape of the new conduit and satisfies the no-slip condition on ∂D, $w\,|_{\partial D} = 0$, $R^2 - r^2 + \varepsilon_1 r^n \sin n\theta = 0$ in polar coordinates (r, θ). The particular values assigned to the mapping parameters (ε_1, n) determine the contour ∂D of the new conduit. As $\varepsilon_1 \to 0$, the base flow w_0 in the conduit with the selected contour is recovered. The mapping allows large deformations of the closed boundary ∂D_o. The values of $0 < \varepsilon_1 < 1, n > 1$ give rise to an almost infinite variety of tube contours, albeit all exhibiting multiple symmetry. For instance, ($\varepsilon_1 = 0.3845$, $n = 3$) and ($\varepsilon_1 = 0.22, n = 4$) correspond approximately to a triangle and a square, respectively. Shapes which are not part of the spectrum naturally, that is, those which cannot be obtained through a single mapping of the domain D_o of the base flow by varying the values assigned to the parameters (ε_1, n), can be generated through a superposition of shapes obtained via two and more separate mappings, $w = w_b + \sum_{j=1}^{N} \varepsilon_j w_{hj}$, $\varepsilon_j < 1$, $\varepsilon_j < \varepsilon_{jc}$. For instance, the teardrop shape commonly used in industry in extruders can be obtained by a judicious superposition of an ellipse and a triangle. The values of ε_1 in each case have to be carefully adjusted to get the particular teardrop shape sought. A given mapping may not yield a closed curve for arbitrarily assigned pair of values for (ε_1, n). For closed curves, the value of ε_1 cannot exceed a critical upper limit ε_{1c} for a given n. The fact that at a cusp the velocity gradient should be zero determines the upper limit:

$$\varepsilon_{1c} = \frac{2}{n}\left(\frac{n-2}{n}\right)^{(n-2)/2}.$$

In theory both ε_1 and n can assume fractional values. In practice, although fractional values are a must for ε_1, admissible closed-form shapes are given only by integer values of n. Assuming that the boundary ∂D_o of the domain D_o of the flow is a circle, a continuous deformation of the circle is observed with increasing values of $\varepsilon_1 > 0$ for fixed integer n up to a limiting closed boundary with n sharp corners or cusps obtained when ε_1 is equal to a critical value ε_{1c}.

The role of rounded corners should be emphasized in determining the axial and secondary fields in axially symmetric cross sections of the type studied by Siginer and Letelier [15]. The mapping they use described in the preceding paragraph allows coming very close to sharp corners with very small radius of curvature. It should be noted that a very small radius of curvature is required to insure analyticity. Under these conditions, the axial and secondary fields are the same as those which correspond to the same cross section with sharp corners except in the very vicinity of the corner. In fact the axial field in the vicinity of the corner is hardly affected, and only the secondary field is slightly different. The effect of the rounded corners was quantified by Hashemabadi and Etemad [160] using the Reiner–Rivlin model to study viscoelastic fluid flow in square ducts. However, their results are only a qualitative assessment of the effect of the roundedness of the corners given that the Reiner–Rivlin CE would at best represent very slight deviations from Newtonian behavior; even then, its predictions are questionable as it does not predict a first normal stress difference. Hashemabadi and Etemad [160] find that for the range $r < 0.1$ of non-dimensional radii of curvatures r in the corner, the mean S_m and maximum S_{max} transverse velocities

$$S_m = \int_{-0.5}^{0.5} \int_{-0.5}^{0.5} \sqrt{u^2 + v^2} \mathrm{d}x\mathrm{d}y$$

$$S_{max} = \mathrm{Max}\left(\sqrt{u^2 + v^2}\right)$$

are almost flat, that is, they change very little with gradually decreasing curvature radii, when the Reynolds number Re, the non-dimensional second normal stress coefficient Ψ_2^*, and the power index n in the representation of the shear-dependent viscosity are set at $Re = 500$, $\Psi_2^* = 0.005$, and $n = 0.8$. In particular, the change in S_m with varying r in the range $r < 0.1$ is very small.

$$\Psi_2^* = \frac{\overline{w}^{(2-n)}\Psi_2}{\eta D_h^{2-n}}, \quad \eta = K\dot{\gamma}^{(n-1)}, \qquad Re = \frac{\rho D_h^n \overline{w}^{(2-n)}}{K}$$

where $\overline{w}, D_h$, $\dot{\gamma}$ and K represent the average axial velocity, the hydraulic diameter, the shear rate, and the consistency index, respectively. However, if Ψ_2^* is doubled to $\Psi_2^* = 0.01$ keeping the Re and n the same, the same claim cannot be made as S_m and S_{max} both increase significantly with decreasing radii. Although the point is not made in [160], it is suspected that the same may be true for $r < 0.01$. If S_m and S_{max} stay flat in a given region of variation of the curvature radii, the strength of the vortices of the secondary field in the cross section is not affected by the varying roundedness of the corners in that region.

However, the mapping used by Siginer and Letelier [15] allows radii much smaller than $r < 0.01$; thus reasonably sharp-looking corners and therefore close to exact strength of the secondary vortices may be obtained. The analytical method presented is capable of predicting the velocity field in tubes with arbitrary cross section. The base flow is the Newtonian field and is obtained at $\boldsymbol{O}(1)$. Field variables are expanded in asymptotic series in terms of the Weissenberg number We. The analysis is not a perturbation of the rest state and does not place any restrictions on the smallness of the driving pressure gradients which can be large and applies to dilute and weakly elastic non-linear viscoelastic fluids (see the discussion on types of perturbation in Sect. 3.3.2). The velocity field is investigated up to and including the third order in We. The Newtonian field in arbitrary contours is obtained, and longitudinal velocity field components due to shear-thinning and to non-linear viscoelastic effects are identified. Third-order analysis shows a further contribution to the longitudinal field driven by first normal stress differences. Secondary flows driven by unbalanced second normal stresses in the cross section manifest themselves as well at this order (Fig. 3.12).

The convected derivatives of the interpolated Maxwell type embedded in the constitutive structure lead to the change of type of vorticity whenever the viscoelastic Mach number $M > 1$ anywhere in the flow field partitioning the field into compatible supercritical (hyperbolic) and subcritical (elliptical) regions with consequences on the heat transfer problem, Siginer and Letelier [161]. The elasticity

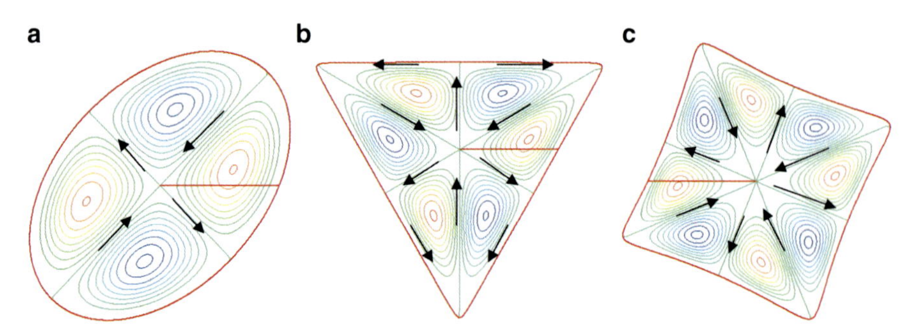

Fig. 3.12 Secondary flow streamlines for an (**a**) elliptical, (**b**) triangular, and (**c**) square cross-section straight tube; $Re = 200$, $We = 0.3$, $\xi = 0.3$ (reprinted from Siginer and Letelier [15] with permission)

number E together with M governs the extent of the hyperbolic region. The size of the supercritical region grows with increasing M. At fixed $M > 1$, the size of the hyperbolic region decreases with increasing E (see Chap. 4 for the definitions of the viscoelastic Mach and Elasticity numbers as well as the hyperbolicity and ellipticity of the vorticity).

The fully developed flow of an Oldroyd type of fluid with three constants in a duct with square cross section has been numerically investigated by Zhang et al. [162] using a finite volume method based on the SIMPLE algorithm. The 3-constant Oldroyd type of CE was previously proposed in relation to the stability study of a torsional flow problem between parallel disks by Phan-Thien and Huilgol [163],

$$\mathbf{T} + p\mathbf{1} = 2\eta_s\mathbf{D} + \mathbf{S}$$

$$\mathbf{S} + \lambda\left(\frac{D\mathbf{S}}{Dt} - \mathbf{LS} - \mathbf{SL}^{\mathrm{T}}\right) = 2\eta_p\left(\mathbf{D} - \mu\lambda\mathbf{D}^2 - \frac{1}{2}\mu\lambda tr\mathbf{D}^2\mathbf{1}\right)$$

where $\mathbf{S}$ is the extra-stress tensor and $\mathbf{L} = (\nabla\mathbf{u})^{\mathrm{T}}$ and $\mathbf{D}$ are the velocity gradient tensor and the symmetric part of $\mathbf{L}$, respectively, with η_s, μ, and λ representing the solvent viscosity, the polymer contributed viscosity, and the material parameters. In a simple shear flow, this model predicts for the first and second normal stresses N_1 and N_2,

$$N_1 = S_{xx} - S_{yy} = 2\lambda\left(\dot{\gamma}\right)^2\eta_p$$

$$N_2 = S_{yy} - S_{zz} = -\frac{1}{2}\mu\lambda\left(\dot{\gamma}\right)^2\eta_p$$

The solution for the secondary flow predicted by this equation is rather interesting because of the very unusual secondary flow patterns not predicted by other CEs. The number of vortices switches from 8 to 16 and back to 8 again during the evolution of the secondary flow structure at constant Reynolds number Re and viscosity ratio η_p/η ($\eta = \eta_s + \eta_p$) with either increasing We at a fixed value of the material constant μ or varying μ at constant We, for example, when $\eta_p/\eta = 0.2$, $Re = 50$, and $\mu = 0.2$, with $3 < We < 5$. Of course these predictions are to stand the test of experiments which have not been undertaken as yet at the time of this writing.

3.6.2 An Industrial Application: Viscous Encapsulation

Stratified immiscible laminar flows have many applications in industry. The energy required to transport heavy viscous crude oil in long pipes can be considerably reduced by the addition of small amounts of water to the crude resulting in reduced pumping costs, Russel and Charles [164] and Charles and Lilleleht [165]. In the fiber industry, stratified flow is used to make bicomponent fibers by coextruding two components in a side-by-side semi-circular configuration to produce unique fiber properties resembling those of natural wools called crimped or wool fibers, Blais et al. [166]. It is common practice in industry to join together in a feedblock multiple layers of several polymers to form and to extrude a single material structured in multiple layers. That practice allows the manufacturer to combine and maximize the desirable properties of multiple polymers into a single structure to enhance the performance properties of the extruded material for a particular application. This process called coextrusion is commonly used to manufacture multilayered blown film, tubing, wire coatings, and other applications. An often encountered problem with co-extruded materials is a phenomenon called *viscous encapsulation* due to the viscosity differences of the layers. It is well documented that in a two-layer configuration, for example, the less viscous layer tends to *encapsulate* the more viscous layer resulting in non-uniform layer thicknesses, Southern and Ballman [167] and Khan and Han [168] to name a few among a flurry of publications in particular in the 1960s and the 1970s on this subject. In a comprehensive experimental and analytical study, the last authors conclude that in conduits with rectangular cross section, the more viscous component tends to push into the less viscous component consistent with experimental observations. Experimentally they establish that the more elastic component tends to wrap around the less elastic component as the normal stress across the interface is not balanced. The component with a greater normal stress will push into the other component with a lower normal stress pointing at the interface. Khan and Han [168] seem to establish that the viscosity difference between the two components predominates over the elasticity ratio in determining the interface shape in stratified two-phase flow. However, this claim is based on their analytical investigation using a second-order constitutive model for either component, thus neglecting secondary flows as the model adopted does not predict a transversal field (see Sect. 3.4). In addition even when viscosities of the layers are well matched, experiments with multilayered systems show that layer thicknesses are non-uniform. This perplexing phenomenon has been clarified by the experiments of Dooley et al. [169] who showed that secondary flows are the physical mechanism behind the non-uniform layer thicknesses in multilayered systems with well-matched viscosities. They conducted

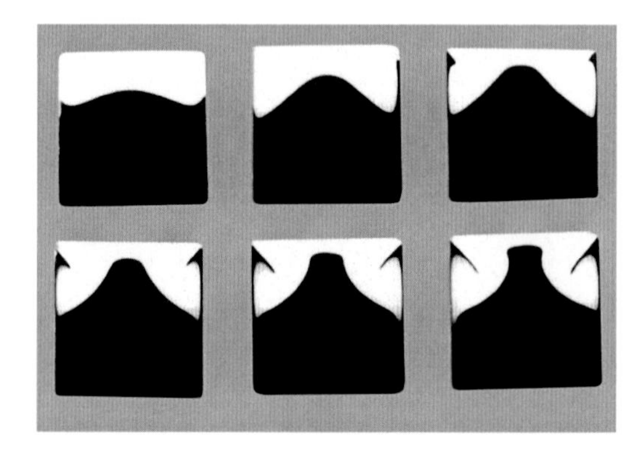

Fig. 3.13 Cross sections of two-layer co-extruded polyethylene structure as it progresses through a rectangular channel at 7.6 cm intervals clockwise (from Dooley et al. [169] with permission)

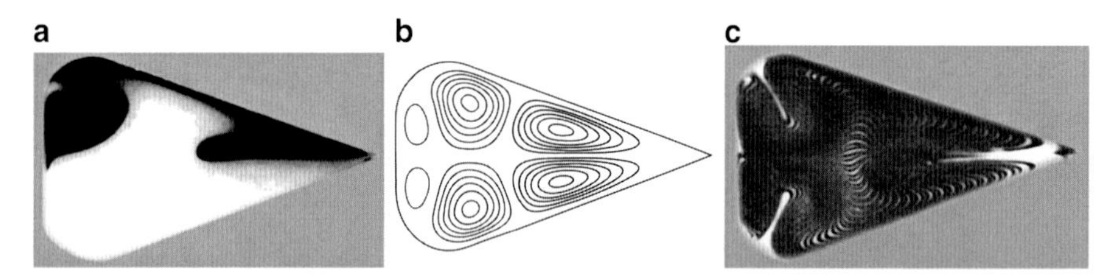

Fig. 3.14 Cross sections of a (**a**) 2-layer and a (**c**) 165-layer coextruded polystyrene structure in a teardrop channel at 50 cm downstream from the feedblock (reprinted from Dooley et al. [169] with permission); (**b**) secondary flow pattern in established steady flow in a teardrop channel

extensive experiments in square, rectangular, teardrop, and circular channels with pigmented layers of the same polymer to eliminate viscosity effects. The material is constructed of alternating black-and-white-pigmented layers which number up to 165. During the flow through the die secondary flows cause drastic rearrangement of the layers which cannot preserve their original horizontal configuration in the beginning of the die. Three resins of increasing elastic properties thus increasing second normal stress differences are studied. The deformation of the interface between layers is more pronounced with the more elastic resins such as polystyrene and polyethylene (Figs. 3.13, 3.14, and 3.15).

The geometry of the die also had a significant impact on the layer rearrangement. The interfaces in the square and teardrop channels showed large deformations, whereas in rectangular sections with relatively large aspect ratios like 4:1, interface deformations are mostly concentrated near the short sides of the rectangle, and over most of the cross section, the interfaces are almost flat in keeping with the distribution of the intensity of secondary vortices. As a check, very little if any interface deformation was observed in a circular channel. In the light of the experimental evidence available, it seems to this author that the claim of Siline and Leonov [156] that the effect of secondary flows can be ignored in most industrial processes is rather premature.

Fig. 3.15 Cross section of a 165-layer co-extruded polystyrene structure in a square channel at 50 cm downstream from the feedblock (reprinted from Dooley et al. [169] with permission)

Dooley and Rudolph [170] experimenting with co-extruding four different polyethylene resin combinations, some with similar and some with dissimilar viscosities, through two different channel geometries (circular and square) to isolate the individual effects of viscosity and elasticity confirmed and extended the above findings. They conclude that when the viscosity difference is large between the layers of polyethylene, viscous encapsulation is a very strong effect in both the circular and square channel with and without secondary flows in the former and the latter, respectively. However, when the viscosity difference is small, the elastic layer rearrangement effect becomes very important for the final interface shape in the square channel due to the strong effect of the secondary flows.

Chapter 4
Transcriticality

Abstract The partitioning of the flow domain of fluids of instantaneous elasticity into two distinct regimes, a relatively high-speed supercritical region away from the boundaries where the viscoelastic Mach number $M \gg 1$ and where the vorticity is governed by a hyperbolic equation and a low-speed region near the boundaries where it is governed by an elliptic equation, is discussed. The smallness of the Reynolds *Re* number is not a sufficient reason for neglecting the inertial non-linearity in the flow of viscoelastic fluids with instantaneous elasticity, those which do not have a Newtonian base viscosity. The pivotal parameter is the product of the Reynolds number and the Weissenberg number $M^2 = Re\ We$, where M is the viscoelastic Mach number, the ratio of a characteristic velocity of the fluid to that of the speed of shear waves. The neglect of the inertial terms may only be justified for small values of the Mach number M.

Keywords Change of type of vorticity • Hyperbolic vorticity • Elliptic vorticity • Viscoelastic Mach number • Instantaneous elasticity • Elasticity number • Wave velocity

The smallness of the Reynolds number is not a sufficient reason for neglecting the inertial non-linearity in the flow of viscoelastic fluids with instantaneous elasticity, those which do not have a Newtonian base viscosity. The pivotal parameter is the product of the Reynolds number and the Weissenberg number $M^2 = ReWe$, where M is the viscoelastic Mach number, the ratio of a characteristic velocity of the fluid to that of the speed of shear waves as Joseph and his coworkers have shown for isothermal flows of fluids with instantaneous elasticity [171]. The neglect of the inertial terms may only be justified for small values of the Mach number. Yoo and Joseph [172] have shown that in the Poiseuille flow in a wavy channel, the vorticity of a UCM (upper convected Maxwell) fluid will change type when the velocity in the center of the channel is larger than a critical value defined by the propagation of the shear waves. Their conclusions are extended by Joseph [171] to Poiseuille flow in longitudinally wavy pipes. The flow domain is partitioned into two distinct regimes. There is a relatively high-speed central region of the channel where vorticity $\boldsymbol{\zeta} = \nabla \times \mathbf{u} = \zeta\,\mathbf{e}_\theta$ is governed by a hyperbolic equation and a low-speed region near the channel walls where it is governed by an elliptic equation. The hyperbolic region is characterized by elastic effects like wave propagation along the

D.A. Siginer, *Developments in the Flow of Complex Fluids in Tubes*,
DOI 10.1007/978-3-319-02426-4_4

characteristics and undamped oscillations of the perturbed vorticity, whereas in the elliptic region the vorticity oscillations are damped and discontinuities if any are smoothed. In the hyperbolic region it is possible that the derivatives of the vorticity are discontinuous across the characteristics if the boundary data is discontinuous. The supercritical and subcritical subdomains are compatible and are determined by a viscoelastic Mach number M defined as the ratio of the unperturbed maximum velocity U to the speed c of the shear waves into the fluid at rest. The Mach number is larger than one in the supercritical hyperbolic region around the center of the channel $M > 1$. The subcritical region grows in size as the elasticity number E increases at a fixed $M > 1$. The vorticity dynamics in the smaller hyperbolic region shows increased elastic behavior at high E. The vorticity in this region shows an oscillatory behavior along the characteristics. The period of the oscillation is governed by M. When the Mach number $M \gg 1$, the thickness of the hyperbolic region is small if the dimensionless elasticity number E is large and vice versa.

The elasticity number is independent of the velocity and is defined as $E = We/Re = \eta\lambda/\rho a^2$, where η, λ, ρ, and a are the zero shear viscosity, the natural relaxation time of the fluid, the density, and the radius of the unperturbed boundary, respectively. This definition of the elasticity number, the ratio of elastic forces and viscous forces, is good for strongly subcritical flows $M \ll 1$. However, for studies of change of type, it is better to use an equivalent definition $E = c^2/(a/\lambda)^2$, the square of the ratio of diffusion velocity to wave velocity. In steady flows the physics is better clarified if $\sqrt{E} = \lambda(\nu/a)/\lambda c$, the ratio of the length influenced by diffusion over a time λ to the length traveled by a wave over the same time, is used. For large values of E, the oscillations are not damped as the centerline is approached. The larger the elasticity number, the larger is the amplitude of the oscillations when M is fixed. There is a rapid damping of vorticity in the supercritical region away from the boundary when $M \gg 1$ if the Weissenberg number $We = M\sqrt{E} \leq O(1)$. The rate of damping of vorticity decreases with increasing We. Flows with high M appear to be more elastic in the sense that damping is suppressed as the relaxation time of the fluid is increased.

Similar criteria apply to the class of non-affine fluids with interpolated Maxwell type of convected derivatives. In his 1990 monograph Joseph develops a complete overview of the field. However, he only considers axisymmetric cases such as pressure-gradient-driven flow of UCM fluids in longitudinally wavy channels and pipes as well as the boundary-driven flow between rotating corrugated cylinders. In all these steady axisymmetric cases, vorticity has only one component $\boldsymbol{\zeta} = \nabla \times \mathbf{u} = \zeta\, \mathbf{e}_\theta$ and the corresponding equations read

$$A\zeta_{,zz} + 2B\zeta_{,rz} + C\zeta_{,rr} = l.o.t.,$$

$$A = M^2u^2 - (We\, S_{rr} + 1), \quad B = M^2uw - We\, S_{rz}, \quad C = M^2w^2 - (We\, S_{zz} + 1),$$

$$A\zeta_{,rr} + 2B\frac{1}{r}\zeta_{,r\theta} + C\frac{1}{r^2}\zeta_{,\theta\theta} = l.o.t.,$$

$$A = M^2u^2 - (We\, S_{rr} + 1), \quad B = M^2uv - We\, S_{r\theta}, \quad C = M^2v^2 - (We\, S_{\theta\theta} + 1),$$

in the pressure-gradient-driven steady flow in a longitudinally wavy tube with $\zeta(r, z)$ and steady flow driven from the boundary between rotating cylinders with $\zeta(r, \theta)$, respectively. In these equations ($l.o.t.$) stands for lower order terms, which includes ζ and its first-order derivatives together with other flow variables and their first- and second-order derivatives. These terms do not have any relevance to the criteria that determines transcriticality. Locally it is possible to determine whether vorticity is hyperbolic or elliptic by studying the characteristic curves of the quasilinear first-order system obtained from the dynamical equations governing the motion. These characteristics are also associated with the vorticity equation. In coordinate-free notation the vorticity equation can be written as

$$\rho\{tr(\mathbf{u} \otimes \mathbf{u})(\nabla \otimes \nabla)\}\boldsymbol{\zeta} - G\,\Delta\boldsymbol{\zeta} - \frac{\xi}{2}\nabla \wedge (\mathbf{S} \bullet \nabla) \wedge \boldsymbol{\zeta} - \frac{2-\xi}{2}tr\{\mathbf{S}(\nabla \otimes \nabla)\}\boldsymbol{\zeta}$$
$$= l.o.t.$$

It can be shown that the coefficients of the leading terms in this equation for the vorticity components ζ_i, $i = 1, 2, 3$ for the UCM model ($\xi = 0$) are all the same implying that the criterion for the change of type for each vorticity component is the same. However, if an interpolated Maxwell convected derivative is used (non-affine model, $\xi \neq 0$), the coefficients are different thereby implying that a different criterion for change of type may be expected for each component ζ_i of the vorticity. In arbitrary non-axisymmetric cross-sectional straight tubes, the vorticity generated at the boundary due to no-slip will in general have three components with different transcriticality criteria and possibly different not necessarily overlapping hyperbolic regions. These observations have relevance to the heat transfer problem with viscoelastic fluids of the interpolated Maxwell type in straight tubes of non-circular cross section. There is strong evidence that the physics underlying the heat transfer problem is based on the change of type of vorticity, Siginer and Letelier [161].

Chapter 5
Quasi-Periodic Flows of Viscoelastic Fluids in Straight Tubes

Abstract The effect of pressure gradient oscillations and longitudinal and transversal boundary oscillations on the flow of non-linearly viscoelastic fluids in circular tubes driven by a constant mean pressure gradient is discussed. Flow enhancement and anomalous flows due to frequency cancellation of superposed boundary waves and resonance like behavior due to the coupling of the viscoelastic and viscous properties leading to drastic enhancement of the instantaneous flow velocities, order of magnitude larger increases at certain frequencies of the driving quasi-periodic pressure gradient oscillating about a zero mean is reviewed. Mean secondary flows of non-linear viscoelastic fluids driven by pulsating pressure gradients in straight tubes of *non-circular* cross section are discussed.

Keywords Oscillating flow • Pulsating flow • Non-linear viscoelastic fluids • Resonance phenomena • Flow enhancement • Frequency cancellation • Anomalous flows • Circular and Non-circular cross-sections

5.1 Pulsating Pressure-Gradient-Driven Flows

It is quite well established both theoretically and experimentally that no non-linear effects manifest themselves in both laminar and turbulent flows of Newtonian liquids under moderate periodic forcing. Time-averaged properties of the pressure-gradient-driven flow of Newtonian fluids are not affected by a pressure gradient pulsating around a mean either zero or non-zero, and steady and pulsating mean velocity profiles are not significantly different except in turbulent flow and then only if the amplitude of the oscillation exceeds the mean pressure gradient by about 20 % at moderate as well as high values of the Reynolds number and at low frequencies. Instantaneous velocity profiles are the same as those corresponding to Poiseuille flow at the instantaneous value of the pressure gradient for low frequency pulsations. Inertial effects start becoming significant at higher frequency values, and flow cannot keep pace with the rapid change in the instantaneous value of the pressure, Ramaprian and Tu [173], Tu and Ramaprian [174], Shemer and Kit [175], and Shemer et al. [176]. Sexl [177] and Lambossy [178] are among the first authors who considered oscillating flow that is zero mean pressure-gradient flow of linear fluids. The more complex problem of pulsating flow in thin-walled elastic

D.A. Siginer, *Developments in the Flow of Complex Fluids in Tubes*,
DOI 10.1007/978-3-319-02426-4_5

and elastico-viscous tubes was investigated by Womersley [179] in the context of blood flow. Flow driven by a pressure gradient oscillating about a non-zero mean has been investigated first by Uchida [180].

In contrast to the Newtonian behavior, the capacity of viscoelastic fluids to store elastic energy and the non-linear dependence of the deformations on the stress field result in dramatic deviations from the Newtonian field of the same density and viscosity. The zero mean pressure-gradient-driven flow of viscoelastic fluids has been considered by Pipkin [181] and Etter and Schowalter [182] who used the Rivlin–Ericksen and the three-constant Oldroyd models, respectively, and by Lanir and Rubin [183] in the context of industrial liquids. The effects of pressure oscillation around a non-zero mean in rigid pipes was studied in chronological order by Walters and Townsend [184] and Barnes et al. [185] followed by the work of Edwards et al. [186], Townsend [187], DaVies et al. [188], Böhme and Nonn [189], Manero and Walters [190], and Phan-Thien [191–193]. Barnes et al. [185], DaVies et al. [188], Böhme and Nonn [189], and Phan-Thien [191, 192] used the Oldroyd, Goddard–Miller, fourth-order fluid, and Maxwell models, respectively, in their analytical investigations, whereas Townsend [187] and Manero and Walters [190] employ a four-constant Oldroyd model together with a finite difference approach which allows the simulation of large amplitudes and frequencies as opposed to the investigations of the remaining authors mentioned above. Blood flow in the microcirculation is pulsatile driven by the pulsations initiated by the heart and can be conceived of as the superposition of a steady shear and an oscillatory shear. Vlastos et al. [194] investigate the viscoelastic behavior of human blood, which at a threshold physiological volume concentration and beyond exhibits non-Newtonian and thixotropic properties. They use polyacrylamide and xanthan solutions to model the rheological properties of human blood. The response of polymer solutions to this type of shear superposition shows qualitative similarities with blood flow.

There is in general agreement that the flow rate increases with the mean gradient, at least for relatively low pressure gradients. But other aspects, for instance, the behavior with increasing frequency at fixed amplitude, were not clarified. Although Huilgol and Phan-Thien [195] indicate that the data of Sundstrom and Kaufman [196] show that flow enhancement decreases with frequency, the experimental data of Sundstrom and Kaufman [196] obtained with a weakly viscoelastic polymeric solution, Natrosol 250-HR, is at best inconclusive and open to interpretation. They do claim, however, that the enhancement effects decrease with increasing frequency and amplitude at fixed mean pressure gradient and reach an asymptotic value for the frequency $\omega \rightarrow \infty$ based on their analytical computations using the inelastic Ellis model, the use of which is only partially justifiable if the period of the oscillation is much greater than the natural response time of the fluid. Barnes et al. [185] on the other hand did observe an increase in the flow enhancement with increasing frequency in the case of polyacrylamide solutions of different concentrations. Their data convincingly shows an increasing trend with the frequency together with an increase in flow enhancement at low values of the mean pressure gradient and a decrease at high values of it at the same frequency. At very low values of the frequency, they show that flow enhancement is governed by the "inelastic" solution, that is, the zero frequency limit, which depends only on the shear-thinning properties of the fluid. Experimentally minimum enhancement effects are obtained at this limit for a given mean pressure gradient, a finding at variance with their analytical computations using Oldroyd model. On the other hand Edwards et al. [186] predict that the change in the flow rate is a constant value independent of the mean pressure gradient, a finding not corroborated by experiments.

The magnitude of the amplitude of the fluctuation is equal to or smaller than the mean pressure gradient in the experiments of Sundstrom and Kaufman [196] and Barnes et al. [185], respectively, leading to data which pertain to superposed equally dominant oscillatory shear and Poiseuille flow and oscillatory flow superposed on a dominant Poiseuille flow, respectively. The simulation of the existing experimental data is a good test to pass for anybody's preferred constitutive equation. The emphasis on all these investigations is on differential constitutive models with the exception of that of Phan-Thien [193] who used the single integral constitutive equation of the generalized Maxwell type with strain-rate memory kernel to analyze the problem for sinusoidal as well as for more general pressure-gradient noises with constant mean values in a statistical sense to qualitatively predict aspects of the frequency data of Barnes et al. [185]. However his work [191, 192] with the differential non-affine network models of Johnson–Segalman and Wagner (see Siginer [73] sections 2.6 and 2.7, respectively, for the Johnson–Segalman and Wagner models), respectively, does not predict the data even qualitatively. The three- and four-constant Oldroyd, Goddard–Miller, fourth-order fluid, corotational Maxwell, non-affine network, and Wagner models (see Siginer [73] for more on these models) do not predict available experimental data for polyacrylamide solutions of different concentrations with the amplitude of the fluctuation smaller than the mean gradient, thus pertaining to oscillatory shear superposed on a dominant Poiseuille flow. Some of these models, Goddard–Miller, Oldroyd, Wagner, and Johnson–Segalman, among them, lead to predictions which indicate the opposite trends shown by the experiments. Frequency dependence of the enhancement in particular is not even qualitatively predicted by any of these models. The studies by Siginer [197–199] who used multiple integral representations of the constitutive structure given by a series of nested integrals with tensor polynomial kernels in terms of the history of the motion $\mathbf{G}(s)$ originally developed by Green and Rivlin [200] and Coleman and Noll [118] predicted many features of the existing data for the first time.

$$\begin{aligned}\mathcal{F}_{ij} &= \int_0^\infty K_{ijkl}(s)G_{kl}(s)\mathrm{d}s + \int_0^\infty\!\!\int_0^\infty K_{ijklmn}(s_1,s_2)G_{kl}(s_1)G_{mn}(s_2)\mathrm{d}s_1\mathrm{d}s_2 \\ &+ \int_0^\infty\!\!\int_0^\infty\!\!\int_0^\infty K_{ijklmntp}(s_1,s_2,s_3)G_{kl}(s_1)G_{mn}(s_2)G_{tp}(s_3)\mathrm{d}s_1\mathrm{d}s_2\mathrm{d}s_3 + \ldots\end{aligned}$$

$$\mathbf{S} = \mathbf{T} + \Phi\mathbf{1} = \mathcal{F}_{s=0}^{\infty}\left[\mathbf{G}(\mathbf{X}, s)\right], \quad \mathbf{G}(t-s) = \mathbf{C}_t(t-s) - \mathbf{1}, \quad \tau \leq t, \quad s = t - \tau$$

The present and past times, the modified pressure field, the total stress, the particle position, and the relative right Cauchy–Green strain tensor are represented by t, τ, Φ, $\mathbf{T}$, $\mathbf{X}$, and $\mathbf{C}_t$ respectively. Mathematically manageable forms of the extra-stress response functional are obtained by linearizing the functional $\mathcal{F}$ by assuming functional differentiability at some deformation history $\mathbf{G}_o$.

$$\mathcal{F}_{s=0}^{\infty}[\mathbf{G}(\mathbf{X}, s)] = \mathcal{F}[\mathbf{G}_o|\mathbf{G}_{oo}] + \delta\mathcal{F}[\mathbf{G}_o|\mathbf{G}_{oo}] + \delta^2\mathcal{F}[\mathbf{G}_o|\mathbf{G}_{oo}, \mathbf{G}_{oo}] + O\left(|\mathbf{G}_{oo}|^3\right)$$
$$\mathbf{G}(\mathbf{X}, s) = \mathbf{G}_o(\mathbf{X}, s) + \mathbf{G}_{oo}(\mathbf{X}, s) \tag{5.1}$$

$\delta\mathcal{F}$ and $\delta^2\mathcal{F}$ represent functional derivatives at $\mathbf{G}_o$, linear and bilinear in $\mathbf{G}_{oo}$, respectively. Functional derivatives $\delta\mathcal{F}$ and $\delta^2\mathcal{F}$ may be assumed to be in integral form if $\mathcal{F}$ is Fréchet differentiable at $\mathbf{G}_o$. However, although it is well known that Riesz theorem justifies the representation of the first Fréchet derivative $\delta\mathcal{F}$ as a single integral with the integrand linear in $\mathbf{G}_{oo}$, there are no representation theorems for higher orders, and the representation of the second $\delta^2\mathcal{F}$ and third $\delta^3\mathcal{F}$ Fréchet derivatives as double and triple integrals, bilinear and trilinear, respectively, in $\mathbf{G}_{oo}$ constitutes only a partially justifiable hypothesis. Siginer [197–199] employed an extension of the algorithm for integral fluids of the Green–Rivlin type originally developed by Joseph [201] by expanding flow variables into formal Taylor series pivoted around the rest state $\mathbf{G}_o$ and expanding the extra-stress into a Fréchet series. The domain of the functional $\mathcal{F}$ is smooth, and the elements of the topological dual are linearized stresses represented by Fréchet derivatives of $\mathcal{F}$ in the weighted Hilbert norm of Coleman and Noll uniformly approximating the functional $\mathcal{F}$ near the base state $\mathbf{G}_o$, the state of rest. It should be noted that canonical representations for $\delta^2\mathcal{F}$ and $\delta^3\mathcal{F}$ are not known when the base state $\mathbf{G}_o$ is a viscometric flow. For more on this constitutive equation as well as others mentioned above, please see Siginer [73, 197–199].

5.2 Flows Driven from the Boundary

Early work for the superposed, parallel steady, and oscillatory shear flows was conducted experimentally by Booij [202, 203] in the late 1960s who experimented with solutions of ethylene–propylene copolymers in decalin superimposing a small oscillatory shear on a steady shear flow in a *Weissenberg* rheogoniometer. He points out that experimental results deviate rather much from those expected from the *Oldroyd* theory. The behavior of K-BKZ fluids (see Siginer [73] section 2.3.2.1 for K-BKZ constitutive equation) was investigated about the same time by Bernstein [204] who also derived relationships between the storage modulus and the first normal stress difference for parallel shearing of the K-BKZ fluids, Bernstein and Fosdick [205]. After Booij in rapid succession in 1971 and 1973, Jones and Walters [206] and Zahorski [207, 208] studied the behavior of the fluids of the integral type and the behavior of simple fluids with superimposed proportional stretch histories, respectively. The presence of higher harmonics in oscillatory testing of non-Newtonian fluids was already demonstrated by the time Goldstein and Showalter [209] and Townsend [187] independently predicted in 1973 possible flow enhancement effects due to the interaction of simple shear with boundary imposed oscillatory shear. Goldstein and Schowalter were also motivated by the possibility of using higher harmonics to characterize rheologically complex fluids through a non-linear analysis. But the Bird–Carreau model they used has serious shortcomings. It does not reduce to the equations of the linear viscoelasticity for small strains in oscillatory shear and does not predict a transition from linear to non-linear viscoelasticity at a finite strain no matter how high the strain rate is. However, Townsend [187] correctly predicted that oscillatory shear imposed from the boundary through the longitudinal oscillations of the pipe wall would yield large flow rate increases. Later Manero and Mena [210] and Manero et al. [211] experimentally investigated the effect of the longitudinal oscillations of the pipe wall on the Poiseuille flow in straight tubes with aqueous solutions of polyacrylamide at various concentrations as working fluids. Their experiments correspond to the parallel superposition of a dominant, primary oscillatory shear onto a secondary simple shear. The consensus is that the change in the volumetric flow rate is a function of the frequency and amplitude of the boundary oscillation at a constant pressure gradient and is a monotonically decreasing function of the pressure gradient at fixed frequency and amplitude. Enhancement reaches a peak at small values of the pressure gradient with a magnitude that can swell up to ten times the flow rate corresponding to the constant pressure-gradient-driven flow.

Simmons [212] is the first researcher to report experimental measurements concerning transversal shearing imposed from the boundary superposed on steady shearing followed by Tanner and Williams [213] who theoretically and experimentally investigated the orthogonal shearing of K-BKZ fluids in the small gap between concentric cylinders with the inner and outer cylinders axially oscillating and steadily rotating, respectively. Their primary goal was to show that K-BKZ fluids are the best suited to describe nearly viscometric flows, but their results were rather inconclusive. Their experiments fall into the category of primary steady shear perturbed by orthogonally superposed small strain oscillatory shear. In an effort to make the data of Manero and Mena [210] and Manero et al. [211] consistent with a theoretical framework, Kazakia and Rivlin [214, 215] investigated the effect of sinusoidal boundary vibration, either longitudinal or transversal, on the Poiseuille flow between parallel plates and in straight round tubes of a slightly non-Newtonian liquid in the sense defined by Langlois and Rivlin [103] and also the effect of the longitudinal and transversal boundary vibration on the steady shear of the Rivlin–Ericksen fluids. Parallel plates are simultaneously and synchronously vibrated in their own plane in a direction either parallel or orthogonal to the constant pressure-gradient-driven flow, and the pipe wall is subjected to longitudinal and rotational vibrations. Kazakia and Rivlin show that inertia is the main driving mechanism and that the fluid has to be shear-thinning for an increase in discharge to occur with both types of fluids. But their analysis cannot account for the magnitude of the increase in volumetric flow rate. They establish for the slightly non-Newtonian liquids they studied that the increase in mass transport is proportional to the square of the frequency of the boundary oscillation for small frequencies and to the square root of the frequency for large frequencies. Their analysis together with that of Böhme and Nonn [189] who used the fluid of grade four in their investigation is open to criticism leveled by stability studies which cast serious doubt on the use of fluids of either complexity n or grade n in *unsteady* flows, Joseph [216], Renardy [217]. Specifically Joseph [216] and Renardy [217] show that these fluids are inadmissible as exact or approximate models in the study of unsteady motions, and that the rest state of these fluids is unstable in the sense of the linearized stability theory. The analysis of Kazakia and Rivlin [214, 215] was extended by Phan-Thien [218, 219] to the case of random longitudinal vibration of the boundary.

For the conclusions of all these investigations to be meaningful, discounting the stability considerations mentioned, the primary shear flow must be sufficiently slow for the slow flow approximation to hold, and the strains and strain rates driven by the oscillatory forcing from the boundary must remain small restricting the boundary oscillation to small amplitudes and frequencies. Phan-Thien [218, 219] also investigated the predictions of the generalized Newtonian, power law, and single integral models with either strain or strain-rate history type of kernels. He concludes that essential features of the phenomenon can be qualitatively described by these models and that a proper choice of the viscosity function may even make quantitative predictions of the flow enhancement possible using the generalized Newtonian (power law) model. But the power-law model points to an infinite enhancement in the limit of infinitesimal pressure gradients, and it falls short of explaining resonance effects observed for certain values of the parameters. Wong and Isayev [220] investigated numerically the orthogonal superposition of small and large amplitude oscillations upon steady shear flow of viscoelastic fluids using the *Leonov* CE. In the case of small amplitude superposition, the predicted results have been found to be in fair agreement with the experimental data at low shear rates and only in qualitative agreement at high shear rates and low frequencies. In the case of large amplitude superposition, orthogonal superposition has less influence on the steady-state shear stresses and the first normal stress difference than the parallel superposition. However, in the orthogonal superposition a more pronounced influence is observed for the second normal stress difference. Kwon and Leonov [221] confirm that the predictions of the *Leonov* model for the orthogonal superposition of small amplitude oscillations on steady flow coincide well enough with the experimental data of Simmons [212], but dispute overall the findings of Wong and Isayev [220].

All the authors previously mentioned maintain that the driving mechanism behind flow enhancement in the vibrating boundary case is the shear-dependent viscosity, and the elasticity of the liquid plays a minor role if at all. However that is not necessarily always the case was shown by Böhme and Voss [222]. Specifically they show that elasticity of the non-linear liquid contributes significantly to the enhancement and is responsible for at least 35 % of it in the case of 1 % polyacrylamide solution in water. Siginer [223, 224] and Siginer and Valenzuela-Rendon [225] using a much more general constitutive model than theirs, a multiple integral representation with tensor polynomial strain history kernels with fading memory, show that indeed elasticity is a major factor determining the enhancement. The algorithm used, which perturbs the rest state does not impose any restrictions on the *rate of deformation*, that is, rapidly varying shear rates can be simulated. But although the strain rates are not small, the strains themselves must remain small for the solution to be valid. Siginer [223, 224] and Siginer and Valenzuela-Rendon [225] work with integral fluid of order three, that is, with Fréchet derivatives up to and including third order $\delta^3\mathcal{F}$ (see Eq. 5.1), and consider flow driven simultaneously by a pressure gradient oscillating around a non-zero mean and boundary waves both longitudinal and orthogonal all represented by finite truncated Fourier series to show that the elasticity of the liquid plays a significant role in enhancement and that the liquid has to be shear-thinning for the enhancement to occur and that flow partially driven from the boundary is an inertial phenomenon. It is important to emphasize that enhancement effects occur independently of the explicit forms of the kernel functions of the integral fluid of order three. Explicit forms of the kernel functions have a significant bearing on the quantitative predictions. They may be introduced as Maxwell functions with multiple relaxation times whose parameters will have to be determined experimentally; however, many other choices are available.

5.3 Resonance Phenomena, Anomalous Flows, and Flow Enhancement

The classical resonance phenomenon of the linear theory of elasticity provides the motivation by direct analogy for the idea that viscoelastic fluids may through a coupling of the viscoelastic and viscous properties show drastic enhancement of the instantaneous flow velocities at certain frequencies of the driving quasi-periodic pressure gradient oscillating about a zero mean. This resonance-like behavior is equally shared by both the linear and non-linear theories of viscoelasticity. That it may lead in the case of non-linear viscoelasticity to order of magnitude increases in the experimentally observed phenomenon of the mean flow rate enhancement in Poiseuille flows driven by pulsating pressure gradients was shown by Andrienko et al. [226]. Zero mean gradient flow of linearly viscoelastic fluids has been studied early enough in the literature, Fredrickson [227]. But, the appearance of the viscoelastic resonance in flows which may be described by the linear theory of viscoelasticity and the consequences of this resonance-like phenomenon for the mean flow enhancement in the range of the non-linear theory of viscoelasticity when the flow is driven by gradients oscillating around a non-zero mean had not been investigated until the work of Andrienko et al. [226]. Various series expansion techniques are not suitable to study the resonance regime because small perturbations of the pressure gradient may cause large changes in velocity magnitudes. Andrienko et al. [226] use the upper convected Maxwell model with a discrete spectrum of relaxation times which for *unidirectional* Poiseuille flows in rigid tubes assumes the following form in cylindrical coordinates (r, ϕ, z),

$$\sigma(r,t) = \sum_{k=1}^{K} \sigma_k(r,t)$$

$$\left(1 + \lambda_k \frac{\partial}{\partial t}\right) \sigma_k(r,t) = \eta_k \frac{\partial w(r,t)}{\partial r}$$

where w is the axial velocity and σ is the rz component of the extra-stress tensor $\mathbf{S}$. λ_k and η_k, $k = 1,\ldots, K$ refer to the kth relaxation time and kth partial viscosity, respectively. These equations reduce to the Oldroyd-B model when $K = 2$ and $\lambda_2 = 0$ and for the Newtonian liquid $K = 1$ and $\lambda = 0$. If a relaxation time in the spectrum $\eta = \sum \eta_k$ is zero, the corresponding partial viscosity is called a Newtonian component of the total viscosity, which acts to smooth the resonance effect. Even a small Newtonian viscosity component can considerably diminish the resonance amplitude, especially for small tube radii, which is interpreted as the result of viscous damping and has been observed in different contexts with a continuous spectrum as well, Siginer [223, 224]. If there are several relaxation times of different order in the spectrum, a role similar to the Newtonian viscosity component is played by the smaller relaxation times. It follows that resonances which correspond to the greater relaxation times in the spectrum are slurred over because of the viscous damping. On the other hand the existence of quite large relaxation times in the spectrum does not cancel or smooth over resonances at the frequencies which correspond to the smaller relaxation times. Resonance curves for a pressure gradient oscillating around a zero mean is shown in Figs. 5.1 and 5.2 for two tube radii which differ by a multiple of 3. The ordinate in these curves is the non-dimensionalized instantaneous velocity and the abscissa is the non-dimensionalized frequency.

Andrienko et al. [226] show that viscoelastic fluids respond with resonance-like behavior to quasi-periodic forcing in tube Poiseuille flows when the constitutive structure is described either by a linear or non-linear stress–strain relationship. The amplitude of the instantaneous velocity may be increased drastically at certain frequencies as a result of the interplay of elastic and viscous properties of the liquid when the pressure gradient oscillates about a zero mean. With a non-linear constitutive structure and the oscillation taking place around a non-zero mean gradient, the increase in the mean flow rate can be enhanced drastically if the frequency of the forcing coincides with the first resonance frequency. The effectiveness of the resonance forced enhancement diminishes gradually at higher resonance frequencies. A Newtonian plateau at high shear rates in a continuous viscosity function or a Newtonian viscosity in a discrete spectrum reduces the enhancement, which is largest at the smallest resonance frequency. At higher resonance frequencies the

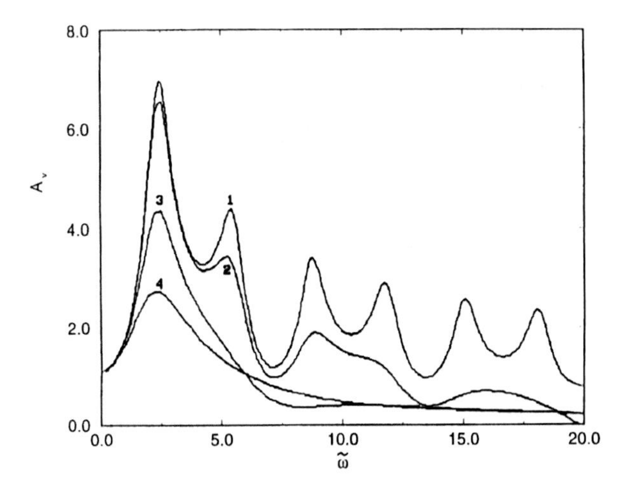

Fig. 5.1 Resonance curves for liquids with two relaxation times $\lambda_1 \neq 0$ and $\lambda_2 = 0$. The partial Newtonian viscosity η_2 corresponding to λ_2 is 0 % of the total viscosity η for the curve labeled "1," and 1 %, 10 %, and 25 % for the curves labeled "2," "3," and "4," respectively. The dimensionless tube radius R and the dimensionless frequency $\tilde{\omega}$ are defined by $R = 0.1\sqrt{\mu\lambda_1}, \mu = \eta/\rho$ and $\tilde{\omega} = \omega\lambda_1$, respectively, with η and ρ representing the total viscosity and density, respectively (reprinted from Andrienko et al. [226] with permission)

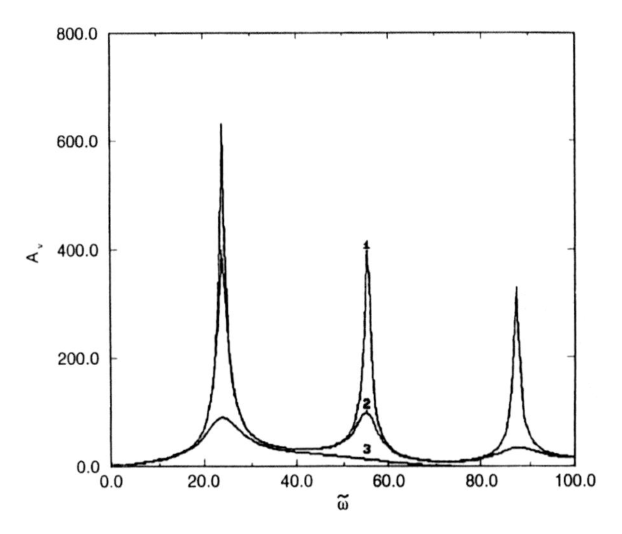

Fig. 5.2 The same as Fig. 5.3 except the tube radius is larger $R = 0.3\sqrt{\mu\lambda_1}, \mu = \eta/\rho$ and the partial Newtonian viscosity is 0 %, 0.1 %, and 1 % for the curves labeled "1," "2," and "3," respectively (reprinted from Andrienko et al. [226] with permission)

decrease in enhancement may be considerable. In general, quite large enhancements can be reached at the resonance frequencies even when the amplitude of the additional oscillatory pressure gradient is very small. The geometry of the tube is a crucial factor in the resonance phenomenon and consequently in the flow rate enhancement effect as well. At larger tube radii the resonance effect gains drastically in amplitude, and much larger enhancements may be obtained.

Steady longitudinal and transversal flows due to frequency cancellation have been studied by Siginer [228]. Specifically it has been shown that when a pressure gradient oscillating around a *zero mean* and longitudinal and rotational sinusoidal waves imposed on the boundary, all of different frequencies and amplitudes, are driving the flow if the frequency of any one forcing component is twice the frequency of another forcing component a steady, mean flow is generated in the direction of the component with twice the frequency. For instance, if the frequency of the pressure gradient oscillation around a zero mean is twice that of either the transversal or longitudinal wall oscillation, a longitudinal steady flow will result. A steady secondary flow will take place if the rotational wave frequency is twice that of either the pressure gradient or the longitudinal wall oscillation due to frequency cancellation.

Flow rate enhancement in the circular tube flow of rheologically complex fluids driven by a pulsating pressure gradient has been studied by Siginer and Valenzuela-Rendon [229] using the constitutive equation of integral fluids of order three. The enhancement is governed by the coupling of the non-linear shear-thinning and elastic properties and the coupling of the linear viscous and elastic properties due to resonance-like phenomena very much along the lines of the findings of Andrienko et al. [226]. For certain values of the constitutive parameters, the couplings become stronger and show resonance-like characteristics. Optimum enhancement is dependent on the shape of the waveform defining the pulsation in the pressure gradient. The squarer the waveform, the larger the enhancement, and the flatter the waveform, the smaller the enhancement. For square waves the enhancement can be as much as 35 %. The existence of a second Newtonian plateau at high oscillatory shear rates (high frequencies) may reduce the enhancement effect substantially as much as 50 % depending on the difference $(\mu - \eta'_\infty)$ where μ and η'_∞ represent the zero shear Newtonian viscosity and the second Newtonian viscosity at high shear rates. The smaller the difference $(\mu - \eta'_\infty)$, the more reduced the enhancement will be as compared to the case $\eta'_\infty = 0$ for any given waveform that defines the pulsation. In pace with the findings of Andrienko et al. [226], the existence of a second Newtonian plateau $\eta'_\infty \neq 0$ confines the enhancement to small values of the frequency and smoothes over any resonance-like effects at higher frequencies in addition to reducing the enhancement substantially. At any frequency and for any waveform, the mean energy required to transport the enhanced flow rate is less than the energy required to transport the same enhanced flow rate with a steady gradient without any superposed oscillations. Energy savings follow a pattern similar to enhancement behavior, that is, the squarer the waveform, the larger the energy savings. For square waves energy savings could be quite substantial as much as 25 %.

That flow enhancement effects along these lines also manifest themselves in two-phase pulsatile, laminar flow of Newtonian/non-Newtonian fluids have been shown by Mai and Davis [230] who investigated the pulsatile, laminar two-phase pipe flow of a Newtonian liquid and a generalized Newtonian fluid or Bingham plastic and potential energy savings that may go along with it. A lighter, less viscous lubricating and immiscible Newtonian fluid lies above a non-Newtonian fluid whose constitutive equation is either a power law or a modified Bingham plastic. The flux of the non-Newtonian fluid in steady flow can be significantly larger than in the corresponding single-phase fully filled flow. Mai and Davis [230] show that such flow rate gains can be further enhanced by introducing an oscillatory component into the applied pressure gradient at a cost of additional power that is significantly less than that corresponding to the fully filled, single-phase pipe flow.

The results of Andrienko et al. [226] obtained in the very late 1990s were confirmed by the recent experimental and analytical research of Casanellas and Ortín [231, 232] who reported analytical studies of the laminar oscillatory flow of Maxwell and Oldroyd-B type of viscoelastic fluids in rectangular and cylindrical cross-sectional tubes and experimental results of the oscillatory flow of a solution of giant micelles in a narrow vertical tube under small driving amplitudes in laminar flow and a large range of driving frequencies. Their theoretical analysis shows that there are resonant velocity peaks under certain conditions and that a very small additive Newtonian solvent contribution is sufficient to largely suppress the resonant behavior thereby confirming the conclusions of Andrienko et al. [216]. Further proof and confirmation is provided by high-resolution particle image velocimetry measurements of the flow field in oscillatory flow of a solution of giant micelles in a meridional plane of a narrow vertical tube in their experiments showing that the velocity magnitude at the tube axis peaks at well-defined resonance frequencies, where the phase lag with the forcing changes abruptly.

Resonance behavior of viscoelastic fluids in Poiseuille flow in the presence of a transversal magnetic field has been investigated by Mohyuddin and Götz [233] who analyzed the velocity enhancement and the resonance behavior both numerically and asymptotically in the case of small pipe radii and provided approximations for the resonance frequencies and achievable velocity enhancements. The importance of resonance phenomena in viscoelastic fluid flows, biofluids in particular, in confined geometries has been demonstrated by Torralba et al. [234] who conducted experiments with a Newtonian fluid (glycerol) and a Maxwellian viscoelastic fluid (CPyCl–NaSal in water) driven by an oscillating pressure gradient in a vertical cylindrical pipe. The frequency range explored has been chosen to include the first three resonance peaks (see Andrienko et al. [226] and Siginer and Valenzuela-Rendon [229]). That under laminar conditions the flow of non-linear rheologically complex fluids driven by oscillating pressure gradients with *relatively large amplitudes* is subject to shear-induced instabilities has been shown by Torralba et al. [235]. They observed in their experiments with Newtonian and viscoelastic shear-thinning fluids that the complex fluid exhibits parallel shear flow regime at low amplitudes but becomes unstable at higher amplitudes with the formation of equally spaced, symmetric vortices along the tube. At even higher amplitudes the vortex structure itself becomes unstable, and complex non-symmetric structures develop. That the instabilities observed are due to the complex rheology of the viscoelastic fluids is evident given that $Re < 10^{-1}$. The amplitudes in their experiments ranged from 0.8 to 2.5 mm and the frequencies from 2.0 to 11.5 Hz.

5.4 Pulsating Flow in Tubes of Non-Circular Cross Sections

Mean secondary flows of viscoelastic fluids driven by pulsating pressure gradients in straight tubes of *non-circular* cross section were investigated for the first time by Letelier et al. [236] and Siginer and Letelier [237]. The constitutive structure of the fading memory fluid defined as a series of nested integrals of particle histories over semi-infinite time domains is perturbed (or equivalently the rest state is perturbed) simultaneously with the domain D_o of the base flow through a novel approach to domain perturbation. The mathematical problem reads as

$$\rho \frac{\mathrm{D}\mathbf{u}}{\mathrm{D}t} = -\nabla\Phi + \nabla \cdot \mathbf{S}, \qquad \nabla \cdot \mathbf{u} = 0 \quad \text{in } D,$$

$$D = \left\{ (r, \theta, z) : 0 \le r \le r\big|_{\partial D}, \quad 0 \le \theta < 2\pi, \quad -\infty < z < \infty \right\}$$

$$\mathbf{u}\left(r\big|_{\partial D}, \theta, z, t \right) = 0, \quad \mathbf{u}(0, \theta, z, t) < \infty$$

$$\Phi_{,z} = -\varepsilon(P + \lambda \sin \omega t), \quad \varepsilon < 1, \qquad \mathbf{u} = u\mathbf{e}_r + v\mathbf{e}_\theta + w\mathbf{e}_z$$

$$u(-\varepsilon) = u(\varepsilon), \qquad v(-\varepsilon) = v(\varepsilon), \qquad w(-\varepsilon) = -w(\varepsilon),$$

with the fading memory multiple integral constitutive structure given as a functional $\mathcal{F}$ of the history of the deformation. The present and past times, the modified pressure field, the mean pressure (isotropic stress), and the total stress are represented by t, τ, Φ, P, and $\mathbf{T}$, respectively. Extra-stress $\mathbf{S}$ is expanded in a Fréchet series pivoted around the rest state,

$$\mathbf{S} = \mathcal{F}[\mathbf{G}(\mathbf{X}; \varepsilon)] = \varepsilon\mathbf{S}^{(1)} + \varepsilon^2\mathbf{S}^{(2)} + \boldsymbol{O}\left(\varepsilon^3\right),$$

and remaining flow variables such as the velocity $\mathbf{u}$ and the modified pressure field Φ are expanded in series in ε pivoted around the rest state,

$$\mathbf{u}(\mathbf{X}, t; \varepsilon) = \varepsilon\mathbf{u}^{(1)} + \varepsilon^2\mathbf{u}^{(2)} + \boldsymbol{O}\left(\varepsilon^3\right)$$

$$\Phi(\mathbf{X}, t; \varepsilon) = \varepsilon\Phi^{(1)} + \varepsilon^2\Phi^{(2)} + \boldsymbol{O}\left(\varepsilon^3\right).$$

The expressions of the first- and second-order Fréchet stresses $\mathbf{S}^{(1)}$ and $\mathbf{S}^{(2)}$ are computed as integrals over semi-infinite time domains in terms of the derivatives of the first Rivlin–Ericksen tensor $\mathbf{A}_1 = 2\mathbf{D}$ where $\mathbf{D}$ is the rate of deformation tensor,

$$\mathbf{S}^{(1)} = \int_0^\infty G(s)\,\mathbf{A}_1^{(1)}(t-s)\mathrm{d}s,$$

$$\mathbf{S}^{(2)} = \int_0^\infty G(s)\mathbf{A}_1^{(2)}(t-s)\mathrm{d}s + \int_0^\infty G(s)\mathbf{L}_1(t-s)\mathrm{d}s$$
$$+ \int_0^\infty \int_0^\infty \gamma(s_1,\, s_2)\mathbf{A}_1^{(1)}(t-s_1)\mathbf{A}_1^{(1)}(t-s_2)\,\mathrm{d}s_1\mathrm{d}s_2,$$

$$\mathbf{A}_1^{(2)}(t-s) = \left.\frac{\partial^2 \mathbf{A}_1[\mathbf{G}(\mathbf{X},\, t-s)]}{\partial \varepsilon^2}\right|_{\varepsilon=0}.$$

The explicit definitions of the material functions $G(s)$ and $\gamma(s_1,s_2)$, the shear and quadratic shear relaxation moduli, respectively, are not required for the general theory. For computations in the case of a particular material, the following explicit expressions introduced by Beavers [238] and previously used by Siginer [198, 224] have been adopted by Letelier et al. [236] and Siginer and Letelier [237],

$$G(s) = \frac{k^k \mu}{\theta^k \Gamma(k)} s^{k-1} e^{-ks/\theta}, \quad \theta = -\frac{\alpha_1}{\mu}, \quad 0 < k \le 1,$$

$$\gamma(s_1, s_2) = \alpha_2 \sum_i^N C_{1i} k_i^2 e^{-k_i(s_1+s_2)}, \qquad \sum_i^N C_{1i} = 1,$$

where α_1, α_2, and k are the first and second Rivlin–Ericksen constants and the power index, θ is the primary relaxation time of the fluid, and k_i represents the inverse of the other relaxation times in the spectrum. The time-dependent flow can be solved, but the motion averaged over a period is more interesting as the oscillations take place over the mean. Thus, the properties of the time-averaged mean problem are studied in [226, 227]. Snapshots of the evolution of the mean secondary vortices following continuous mapping of the circle onto gradually evolving shapes culminating in an equilateral triangle are displayed in Fig. 5.3.

The driving mechanism behind the secondary vortices is the distribution of the second normal stresses in the cross section. The absolute value of the second normal stress increases monotonically along the median starting from the center of the triangle where it is zero and reaches a maximum at the wall. Along the wall toward the corner its intensity decreases and exhibits a minimum at the corner. The absolute value of the second normal stress difference increases along the diagonal from the corner toward the center, reaches a local minimum somewhere around midway between the corner and the center, and thereafter decreases to a zero value at the center. This explains the feeding mechanism of the vortices. Particles are pushed along the wall toward the corner where resistance is lower displacing in turn the particles already in residence at the corner which move under the influence of the increasing normal stress differences toward the center along the diagonal.

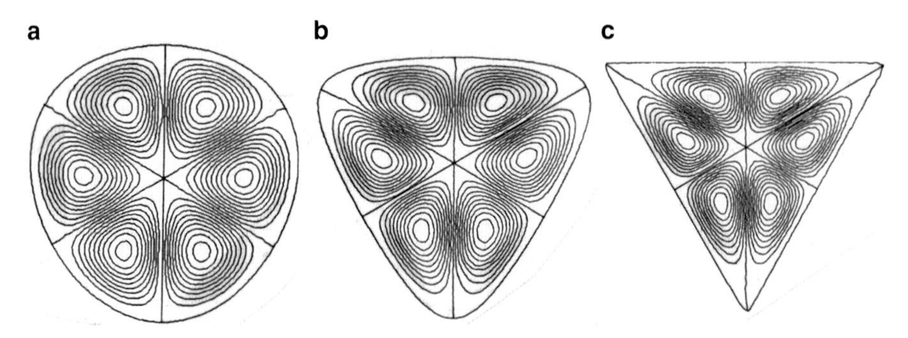

Fig. 5.3 Dimensionless snapshots of the evolution of the mean secondary flow field at various values of ε_1 and fixed $k = 0.5$ as the circular tube is continuously deformed by varying ε_1 to culminate in the triangular tube ($n = 3$, $\varepsilon_1 = 0.845$). The amplitude λ of the oscillation of the pressure gradient is of the same order of magnitude as the mean gradient P, $\lambda/P = 1$. Fluid properties and other data used are given as $\omega = 10$ rad/s, $\rho = 0.89$ g/cm^3, $\mu = 200$ poise, $\alpha_1 = -50$ g/cm, and $R = 3$ cm. $|\Delta\Psi_m| = 0.1 \times |\Psi_m|_{max}$ defines the constant dimensionless increment between streamlines counting down from the maximum value for the mean stream function $|\Psi_m|_{max}$ at the center of each cell. $|\Delta\Psi_m|$ does not represent the jump between the wall where $\Psi = 0$ and the closest streamline contour shown in the figure. The dimensionless stream function Ψ^*_m is defined as $\Psi^*_m = \Psi_m/\Psi_0$ with $\Psi_0 = \frac{PR^3}{4\mu}$. Particles in each vortex close to the wall move towards the corner. (**a**) $n = 3$, $\varepsilon_1 = 0.0769$, $|\Psi_m|_{max} = 2.2 \times 10^{-6}$ (**b**) $n = 3$, $\varepsilon_1 = 0.3076$, $|\Psi_m|_{max} = 1.2 \times 10^{-4}$ (**c**) $n = 3$, $\varepsilon_1 = 0.3845$, $|\Psi_m|_{max} = 4.5 \times 10^{-4}$ (reprinted from Siginer and Letelier [237] with permission)

Chapter 6
Transversal Flow Field of Particle-Laden Linear Fluids

Abstract Mean secondary flows in straight tubes of non-circular cross section turbulent driven of Newtonian fluids by constant pressure gradients are discussed in their historical context as well as in terms of the most recent findings. The fundamental issues and their impact on industrial processes, in particular on processes involving particle laden flows are reviewed. Similarities with the driving mechanism of secondary laminar flows of viscoelastic fluids, criteria for the existence of secondary flows, and general classification and closure approximations for homogeneous and wall-bounded flows are discussed.

The rheology of dilute, semi-dilute, and concentrated non-Brownian suspensions is reviewed. Computing shear viscosity in different concentration regimes and recent progress in determining the normal stress functions of semi-dilute and concentrated non-colloidal suspensions are summarized. Macroscopic constitutive models for suspension flow, shear-induced and stress-induced particle migration, applications to Stokesian dynamics simulations (SDS), and efforts to improve the predictions through SDS both in unbounded and bounded flows are discussed together with challenges in shear-driven migration of non-colloidal concentrated suspensions. The complex nature and sometimes contradictory behavior reported in the literature make it challenging to construct a theoretical model. Efforts to understand the motion of particles in viscoelastic suspending media are summarized and recent research on secondary field in Poiseuille flow of shear-driven migration of suspensions is discussed together with secondary field in single-phase and multiphase turbulent flow of suspensions in tubes.

Keywords Newtonian fluids • Mean secondary field • Turbulent flow • Reynolds stresses • Non-circular tube cross-sections • Secondary flows of the first and second kind • Anisotropic eddy viscosity • Closure approximations • Homogeneous turbulence • Wall-bounded turbulence • Non-Brownian suspensions • Non-colloidal suspensions • Dilute, semi-dilute, and concentrated suspensions • Stokesian dynamics simulations • Shear-induced migration • Stress-induced migration • Shear viscosity of suspensions • Normal stresses of suspensions • Secondary field • Poiseuille flow of suspensions • Secondary field in single-phase turbulent suspension flow • Secondary field in multiphase turbulent suspension flow

D.A. Siginer, *Developments in the Flow of Complex Fluids in Tubes*,
DOI 10.1007/978-3-319-02426-4_6

The structure and the underlying mechanics of the secondary flows of particle-laden Newtonian fluids cannot be grasped unless the transversal field in turbulent tube flow of homogeneous linear fluids is firmly understood. The latest developments in the latter are reviewed first in their historical context.

6.1 Mean Secondary Field in Single-Phase Turbulent Flow

The mean secondary flow in straight tubes of non-circular cross section, a rectangular duct in this case, in single-phase high-Reynolds-number-incompressible turbulent flows was detected for the first time experimentally in the form of two counter-rotating pairs of vortices in each corner of the duct by Nikuradse [99]. He used flow visualization techniques, primarily dye injection, in an attempt to explain the cause of the bulging of the mean turbulent axial velocity contours toward the corner which he had observed in his earlier experiments in several closed geometries of non-circular cross section as well as an open channel of rectangular cross section described in his thesis published in 1926. Specifically in turbulent flow in straight tubes of square cross section, lines of constant velocity are displaced toward the corners and away from the midpoint of the walls compared to those in laminar flow, Nikuradse [239]. Prandtl [240, 241] in a stroke of insight suggested that this shift was due to secondary flows, but he did not provide a rigorous proof to follow up his insight. Einstein and Li [242] in 1958 showed rigorously for the first time that the *gradients of the Reynolds stresses* introduce an *axial mean vorticity* ζ_z which gives rise to the secondary motion in single-phase turbulent flow in straight conduits,

$$\zeta_z = \frac{\partial^2}{\partial x \partial y}\left(\overline{u'u'} - \overline{v'v'}\right) - \left(\frac{\partial^2}{\partial x^2} - \frac{\partial^2}{\partial y^2}\right)\overline{u'v'} \tag{6.1}$$

where u' and v' denote the velocity fluctuations in the cross section and $\overline{u'u'}, \overline{v'v'}$, $\overline{u'v'}$ the Reynolds stresses. z is the axial direction, x the horizontal direction, and y the vertical direction in the cross section. The over bar indicates the time average. The first qualitatively predictive calculations of secondary flows in straight non-circular ducts had to wait until the early 1980s when it was recognized that a model using an isotropic eddy viscosity for calculating the turbulent stresses in the streamwise vorticity equation does not produce any secondary motion at all and that more refined modeling of these stresses was required.

The impact of the secondary flows on engineering calculations is particularly important as turbulent flows in ducts of non-circular cross section are often encountered in engineering practice. Some examples are flows in heat exchangers, ventilation and air-conditioning systems, nuclear reactors, impellers, blade passages, aircraft intakes, and turbomachinery. If neglected, significant errors may be introduced in the design as secondary flows lead to additional friction losses and can shift the location of the maximum momentum transport from the duct centerline. The secondary velocity depends on cross-sectional coordinates alone and therefore is independent of end effects. It is only of the order of 1–3 % of the streamwise bulk velocity, but by transporting high-momentum fluid toward the corners, it distorts substantially the cross-sectional equal axial velocity lines, and specifically it causes a bulging of the velocity contours toward the corners with important consequences such as considerable friction losses. The need for turbulence models that can reliably predict the secondary flows that may occur in engineering applications is of paramount importance.

In the early days, the magnitude of the secondary velocities could not be measured accurately due to the imprecision of the available yaw meters, which introduced significant errors into the measurements when used in flow regions with mean velocity gradients. The advent of the hot-wire anemometers in the late 1950s and early 1960s made these measurements possible. A significant contribution in the early 1960s by Brundrett and Baines [243] analyzed the mean vorticity transport equation to examine the structure of turbulent secondary flow and presented experimental data obtained using hot-wire anemometry for the Reynolds stresses in a rectangular pipe. They showed that it was gradients of Reynolds stresses in the plane of the cross section that gave rise to a source of streamwise vorticity. Their work includes hot-wire measurements of all six components of the Reynolds stress. From these data, they deduced that, in rectangular ducts (and with axes chosen parallel to the sides), it was predominantly the normal stress gradients [the first term in Eq. (6.1)] which generated the velocities in the plane of the cross section. The experiments of Perkins [244] conducted under developing flow conditions near one corner of a square duct lead to the conclusion that both the first and second terms on the *rhs* of Eq. (6.1) are of equal importance. The first term can be interpreted as a generating term and the second as a transport term through diffusion. These findings were confirmed in the 1990s by other researchers through direct numerical simulations (DNS) of Navier–Stokes equations in a square duct, Huser [245]. Before them, Hoagland [246] in 1960 had already determined in his PhD dissertation that the flow pattern due to secondary flows was exactly that predicted by Prandtl [240, 241], but he did not investigate the cause of it. His experiments involved measurements of the complete 3-dimensional mean velocity distribution and local wall shear stress distribution for turbulent flow in a square pipe using hot-wire anemometry. The difficulty with the secondary flow velocity measurements stems from the magnitude of the latter, which is at most a few percent of the primary axial velocity. This means that any small distortions of the flow pattern caused by the measuring probe could have an appreciable effect on the secondary velocities that were deduced. The accuracy of Hoagland's data was diminished further by the use of a coarse device for measuring the small angle made by the total velocity vector with the duct axis. Despite the comparative imprecision of the measurements, however, Hoagland's work remains a major contribution. In the same vein, Perkins [244] pointed out that the results of Brundrett and Baines [243] were not reliable. The hot-wire procedure they used involved rotating a single-wire sensor to eight different orientations at each point in the flow and getting the final measurement for each of the Reynolds stress terms from four r.m.s. voltage readings may have introduced a 100 % error in the measurements. In spite of their perceived and inherent shortcomings, Hoagland's and Brundrett and Baines' data stand out as the first measurements of secondary flows. They proved to be qualitatively correct and led the way for further analytical developments.

This was followed up by the study of Leutheusser [247] of the flow in square and rectangular ducts, which showed that the inner law provided a good description of the flow near the wall, but away from the wall the outer law failed to describe the velocity distribution. Extensive measurements of turbulent air flow under steady, incompressible, fully developed conditions in both square and rectangular channels, which included the effect of the Reynolds number on the structure of the secondary flow and the directional characteristics of the local wall shear stresses, were presented by Gessner and Jones [248]. Their results indicate that secondary flow velocities, when non-dimensionalized with either the bulk velocity or the axial mean flow velocity at the channel centerline, decrease with increasing Reynolds numbers, and the greatest skewness of local wall shear stress vectors occurs in the near vicinity of corners where secondary flow velocity is maximum. They show through experimental evaluation of terms in a momentum balance along a typical secondary flow streamline that secondary flow is the result of small differences in magnitude of opposing forces exerted by the Reynolds stresses and static pressure gradients in planes normal to the axial flow direction.

Evidence from carefully done experiments, using laser Doppler anemometry in later publications, exists in the more recent literature together with theoretical investigations, Launder and Ying [249], Hinze [250], Demuren and Rodi [251], and Nagata et al. [252], to establish that flow in non-circular cross sections takes place along the bisector toward the corner. The number of cells in a square cross section, for example, will be the same as the number of cells in the same cross section generated in the laminar flow of a non-linear polymeric fluid by the second normal stress difference (see Sect. 3.4), but the circulation direction of the vortices would be the opposite. In Newtonian turbulent flows, it is clockwise and in laminar flows of viscoelastic fluids it is counterclockwise, or in other words in the flow of a Newtonian liquid, the flow will be toward the corner whereas a viscoelastic fluid will flow away from the corner toward the center. Research on turbulent secondary flows up to 1987 has been summarized in the review article by Bradshaw [253]. However, Bradshaw concentrates on external flows with very little focus if at all on internal flows.

It is customary in fluid mechanics literature to differentiate between secondary flows driven by turbulence itself, specifically by the anisotropy of the Reynolds stresses, Brundrett and Baines [243], Speziale [97], and secondary flows due to the tilting and stretching of vortex rings promoted by pressure gradients, buoyancy forces in the tube cross section as may happen in curved tubes, or any kind of forcing in the tube cross section such as that due to the forcing of particles in particle-laden flows (see Sect. 6.2). The former is termed Prandtl's secondary flows of the second kind and the latter secondary flow of the first kind. Both are of importance to industrial operations as secondary flows will significantly increase heat and mass transfer within the duct cross section. Secondary flows of the second kind are typically one order of magnitude smaller than the mean streamwise velocity, whereas the magnitude of the secondary flows of the first kind generated by non-conservative body forces perpendicular to the primary motion can be up to 30 % of the mean axial velocity.

We know now that the secondary field of the second kind in Newtonian turbulent flow is determined by the projection of the Reynolds stress tensor onto the pipe cross section, and the direction of the secondary flow is dictated by the gradients of the normal Reynolds stresses $\partial\tau_{rr}/\partial r$ and $\partial\tau_{\theta\theta}/\partial\theta$ with (r,θ) representing radial and azimuthal coordinates in the tube cross section. The Reynolds stress tensor will become anisotropic due to any factor that will break the axisymmetry of the flow. In non-circular tubes, the anisotropy may be generated by the geometry, whereas in circular tubes, it may be caused by the non-uniformity of the boundary conditions such as non-uniform distribution of wall roughness on the boundary, Hinze [250]. For instance, in rectangular cross sections, the transverse Reynolds stresses are anisotropic near the corners of the tube, and this anisotropy generates a pair of counter-rotating vortices near each corner.

The experimental data and theoretical deductions of Hinze [250] are remarkable. He investigated turbulent air flow in a rectangular duct of large aspect ratio with asymmetric roughness and found that in fully developed flow in regions of high turbulence production, where turbulent kinetic energy production exceeds viscous dissipation, there must be an interchange of fluid from wall zones to the core. In wall turbulence usually in regions of high wall shear stresses or of high wall shear stress effects (as in corner regions), the local turbulence production is relatively higher than the local viscous dissipation, while it is the other way around in regions of low wall shear stress effects. The secondary flow will always be toward the corner along the bisectrix in the case of non-circular ducts or along the normal toward a rough wall. Hinze [250] used the full mechanical energy balance equation,

$$\frac{1}{2}\left[\bar{u}_k\frac{\partial\overline{u_i'u_i'}}{\partial x_k}+\frac{\partial}{\partial x_k}\overline{u_k'\left(u_i'u_i'+2\frac{p}{\rho}\right)}\right]=-\overline{u_k'u_1'}\frac{\partial\bar{u}_1}{\partial x_k}-\left(\overline{u_2'u_2'}-\overline{u_3'u_3'}\right)\frac{\partial\bar{u}_2}{\partial x_2}$$
$$+\nu\frac{\partial}{\partial x_k}\overline{\left[u_i'\left(\frac{\partial u_i'}{\partial x_k}+\frac{\partial u_k'}{\partial x_i}\right)\right]}-\nu\overline{\left(\frac{\partial u_i'}{\partial x_j}+\frac{\partial u_j'}{\partial x_i}\right)\frac{\partial u_i'}{\partial x_j}},\quad k=2,3;\quad i,j=1,2,3 \tag{6.2}$$

and measured all the terms in an appropriately reduced equation to deduce his conclusions. In writing (Eq. 6.2), mass conservation

$$\frac{\partial\bar{u}_k}{\partial x_k}=0\quad k=2,3$$

is taken into account. In Eq. (6.2), the mean velocities are represented by $\bar{\mathbf{u}}$ and the fluctuating velocities by $\overline{\mathbf{u}'}$. Measurements were taken in the plane of symmetry

$x_3 = 0$ at a Reynolds number of 1.5×10^5 in a uniform rectangular cross section of $0.45 \times 0.09\ \text{m}^2$ at the very end of a 19 m long tube with long sides in the horizontal plane under fully developed conditions. The long bottom wall was artificially roughened with glued plastic grains except for a central strip of 0.1 m which remained smooth. The average roughness height so obtained was 0.004 m. The remaining walls were smooth.

In a landmark paper, Speziale [97] expressed the results of Brundrett and Baines [243] and Gessner and Jones [248] in a compact and more precise form and showed that secondary flow occurs when the axial mean velocity gives rise to a non-zero normal Reynolds stress difference on planes perpendicular to the axial flow direction. He went on to show that the popular isotropic k-ε and k-l models of turbulence have no driving mechanism for secondary flows in tubes of non-circular cross section, while second-order closure models do. In the absence of body forces, the Reynolds equation for the Reynolds stress tensor $\boldsymbol{\tau}$ reads as

$$\rho\left(\frac{\partial \bar{u}_k}{\partial t} + \bar{u}_l \frac{\partial \bar{u}_k}{\partial x_l}\right) = -\frac{\partial \bar{p}}{\partial x_k} + \mu\, \Delta \bar{u}_k + \frac{\partial \tau_{lk}}{\partial x_l}$$

The mean field $\bar{\mathbf{u}}$ cannot be computed from this system of equations unless a turbulent closure relationship between the Reynolds stress tensor $\boldsymbol{\tau}$ and the mean velocity field $\bar{\mathbf{u}}$ is given, that is, the system is not closed. The mean vorticity $\bar{\boldsymbol{\zeta}}$ transport equation is derived by taking the curl of this equation,

$$\rho\left(\frac{\partial \bar{\zeta}_k}{\partial t} + \bar{u}_l \frac{\partial \bar{\zeta}_k}{\partial x_l}\right) = \rho \bar{\zeta}_l \frac{\partial \bar{u}_k}{\partial x_l} + \varepsilon_{klm} \frac{\partial^2 \tau_{nm}}{\partial x_l x_n} + \mu\, \Delta \bar{\zeta}_k$$

By considering the axial component of the vorticity transport equation, Speziale [97] shows that the development of secondary flow is directly tied to the presence of a non-zero axial mean vorticity $\bar{\zeta}_z$, and that the axial vorticity source term given by

$$\frac{\partial^2 \left(\tau_{yy} - \tau_{xx}\right)}{\partial x \partial y} + \frac{\partial^2 \tau_{xy}}{\partial x^2} - \frac{\partial^2 \tau_{xy}}{\partial y^2} \tag{6.3}$$

is the direct cause of the secondary flow. He further refines his argument to state that if the axial mean velocity $\bar{u}_z$ gives rise to a non-zero normal Reynolds stress difference

$$\tau_{yy} - \tau_{xx} \neq 0 \tag{6.4}$$

secondary flow will occur. He thus states a sufficient condition for the development of secondary flows: *Secondary flow will develop in pipes of non-circular cross section if the axial mean velocity gives rise to any non-zero difference in the normal Reynolds stresses on planes perpendicular to the axial flow direction.*

Another way to state this important result is the following: in fully developed incompressible turbulent flow of Newtonian fluids, the two-dimensional mean transversal velocity $\mathbf{u}$ in the cross section of the tube is determined by the projections onto the cross section of the averaged two-dimensional mass conservation and Navier–Stokes equations,

$$\rho \frac{\mathrm{D}\mathbf{u}}{\mathrm{D}t} = -\nabla \widetilde{p} + \eta \Delta \mathbf{u} + \nabla \bullet \widetilde{\boldsymbol{\tau}}, \qquad \nabla \bullet \mathbf{u} = 0 \tag{6.5}$$

where ρ and η are the density and the viscosity, respectively, and $\nabla \widetilde{p}$ and $\widetilde{\boldsymbol{\tau}}$ are the projections onto the tube cross section of the gradient of the mean pressure and the Reynolds stress tensor. The divergence $\nabla \bullet \widetilde{\boldsymbol{\tau}}$ of the projection $\widetilde{\boldsymbol{\tau}}$ onto the cross-sectional plane of the tube drives the secondary flow in the cross section. Secondary flows in non-circular ducts are accompanied by a longitudinal component of vorticity. The transport equation for the vorticity ζ of $\mathbf{u}$ can be obtained by taking the curl of the linear momentum balance in the cross section

$$\rho \frac{\mathrm{D}\zeta}{\mathrm{D}t} = \eta \Delta \zeta + \nabla \wedge \nabla \bullet \widetilde{\boldsymbol{\tau}} \tag{6.6}$$

As the source term for the vorticity ζ is $\nabla \wedge \nabla \bullet \widetilde{\boldsymbol{\tau}}$, a necessary and sufficient condition for the existence of secondary flows is $\nabla \wedge \nabla \bullet \widetilde{\boldsymbol{\tau}} \neq 0$. The velocity at a location $\boldsymbol{x}_1$ in the cross section is given by the generalized Helmholtz decomposition,

$$\mathbf{u}(\boldsymbol{x}_1) = (2\pi)^{-1} \int_A \frac{\boldsymbol{r}(\boldsymbol{x}_1, \boldsymbol{x}_2)}{\boldsymbol{r}^2(\boldsymbol{x}_1, \boldsymbol{x}_2)} \wedge \zeta(\boldsymbol{x}_2) \mathbf{e}_z dA(\boldsymbol{x}_2)$$

where $\boldsymbol{r}(\boldsymbol{x}_1, \boldsymbol{x}_2)$ is the distance vector between the position vectors $\boldsymbol{x}_1$ and $\boldsymbol{x}_2$ in the cross section of area A, $\boldsymbol{r}(\boldsymbol{x}_1, \boldsymbol{x}_2) = \boldsymbol{x}_2 - \boldsymbol{x}_1$. It is of some importance to note that Brundrett and Baines [243] had deduced the result (Eq. 6.3). But they used it to determine the lines of symmetry of the secondary flow and to establish that the secondary flow consists of eight vortices in rectangular cross sections. Speziale [97] took the step to establish the necessary and sufficient criterion (Eq. 6.4) for the existence of the secondary flow.

6.1.1 *Similarity with the Driving Mechanism of Secondary Flows of Viscoelastic Fluids*

Speziale [97] demonstrates that the popular k-ε turbulence model, a direct generalization of the eddy-viscosity concept,

$$\tau_{kl} = \frac{1}{3}\delta_{kl} tr\boldsymbol{\tau} + C\rho l\sqrt{k}\left(\frac{\partial \overline{u}_k}{\partial x_l} + \frac{\partial \overline{u}_l}{\partial x_k}\right), \quad k = -\frac{1}{2\rho} tr\boldsymbol{\tau}, \quad l = \frac{k^{3/2}}{\varepsilon}$$

where k, C, l, and ε represent the turbulent kinetic energy, a dimensionless constant, the length scale of turbulence, and the turbulent dissipation rate, respectively, does not have the ability to simulate the secondary flows of linear fluids predicting erroneously a unidirectional mean turbulent flow, a deficiency that arises from the use of an eddy-viscosity model based on the Boussinesq hypothesis (see below). The turbulent kinetic energy k is determined from a version of the turbulent kinetic energy transport equation of Mellor and Herring [254],

$$\frac{\mathrm{D}k}{\mathrm{D}t} = -\frac{1}{\rho}\frac{\partial}{\partial x_k}\left(\overline{p' u'_k}\right) - \frac{1}{2}\frac{\partial}{\partial x_k}\left(\overline{u'_k u'_l u'_l}\right) + \frac{1}{\rho}\tau_{kl}\frac{\partial \overline{u}_k}{\partial x_l} + \nu\,\Delta k - \varepsilon$$

$$\varepsilon = \nu\overline{\frac{\partial u'_k}{\partial x_l}\frac{\partial u'_k}{\partial x_l}}$$

where $\overline{p'}$, ν, $\overline{\mathbf{u}'}$, and $\overline{\mathbf{u}}$ are the fluctuating pressure, the kinematic viscosity, and the fluctuating and mean velocity fields, respectively. An overbar indicates steady turbulence time averages. The scalar length scale l is computed differently for the k-ε and k-l models. With the k-ε model $l = \frac{k^{3/2}}{\varepsilon}$ and ε is determined from a version of the scalar dissipation rate equation of Launder et al. [255].

Speziale [97] shows that with the k-ε and k-l models of turbulence, the axial mean velocity $\overline{u}_z$ gives rise to a zero axial vorticity source term, therefore proving that these models cannot predict secondary motions. In fact, any eddy-viscosity model based on the Boussinesq hypothesis

$$\tau_{ij} = \frac{2}{3}k\delta_{ij} - \nu_{\mathrm{T}}\left(\frac{\partial \overline{u}_i}{\partial x_j} + \frac{\partial \overline{u}_j}{\partial x_i}\right)$$

where $\tau_{ij} = \overline{u'_i u'_j}$ and $k = \frac{1}{2}\tau_{ii}$ represent the Reynolds stress tensor and the turbulent kinetic energy, respectively, and ν_{T}, $\overline{\mathbf{u}}$ stand for the eddy viscosity and the mean velocity, is incapable of predicting secondary flows in straight non-circular ducts since the axial velocity will give rise to a vanishing normal Reynolds stress difference. This leads to the conclusion that *anisotropic* eddy-viscosity models, which include *non-linear* strain-dependent terms, constitute the simplest level of *Reynolds stress closure* with the capacity to predict secondary flows in straight *non-circular ducts.*

Speziale derived an anisotropic model purely on continuum mechanical grounds, assuming that the effect of turbulence on the mean flow can be represented by a non-Newtonian relation between stress and strain rate. Making this analogy between turbulence and a non-Newtonian fluid is attractive in the context of flow in a non-circular duct because of the secondary flows that accompany the laminar flow of a non-Newtonian fluid in a non-circular duct. In an effort to remedy the defect in the k-ε and k-l models, Speziale introduced non-linearity into the anisotropic part $\boldsymbol{\tau}^A$ of the Reynolds stress tensor and expressed it as a non-linear function of the mean velocity gradients $\boldsymbol{\tau}^A = \boldsymbol{\tau}^A(\nabla\overline{\mathbf{u}}, k, l, \rho) = \boldsymbol{\tau} - \frac{1}{3}\mathbf{1}tr\boldsymbol{\tau}$. Then invariance considerations yield the following equation:

$$\boldsymbol{\tau}^A = \alpha_0\left(k, l, \rho, tr\overline{\mathbf{D}}^2, tr\overline{\mathbf{D}}^3\right)\mathbf{1} + \alpha_1\left(k, l, \rho, tr\overline{\mathbf{D}}^2, tr\overline{\mathbf{D}}^3\right)\overline{\mathbf{D}} + \alpha_2\left(k, l, \rho, tr\overline{\mathbf{D}}^2, tr\overline{\mathbf{D}}^3\right)\overline{\mathbf{D}}\,\overline{\mathbf{D}} \tag{6.7}$$

where $\overline{\mathbf{D}}$, k, l represent the average rate of deformation tensor, the turbulent kinetic energy, and the length scale of turbulence. But this is the generalized version of the well-known "Reiner–Rivlin" equation, which is known in non-Newtonian fluid mechanics to generate secondary flows. Now this generalized version predicts a non-zero normal Reynolds stress difference $(\tau_{yy} - \tau_{xx}) \neq 0$ and therefore predicts secondary flows in the turbulent flow of linear fluids in tubes of non-circular cross section. The normal Reynolds stress difference $(\tau_{yy} - \tau_{xx})$ is a function of the α_2 very much like the normal stress difference is in the case of viscoelastic fluids if Reiner–Rivlin equation is used to characterize their constitutive structure. Although Eq. (6.7) predicts qualitatively the secondary flow pattern in rectangular and elliptical cross sections in turbulent channel flow, it prescribes $\tau_{yy} = \tau_{zz}$ contradicting experimental evidence, Hinze [256], thus casting serious doubt on the ability of any generalization of the k-ε and k-l models to predict *quantitatively* turbulent secondary flows. However, anisotropic eddy-viscosity models have been introduced with success, some more than others. We mention three: the non-linear k-ε model of Speziale [257]; two-scale direct interaction approximation (DIA) model of Yoshizawa [258, 259], Shimomura and Yoshizawa [260], and Nisizima and Yoshizawa [261] who proposed an improvement of the eddy-viscosity representation for Reynolds stress based on the statistical viewpoint in which the Reynolds stresses are quadratic functions of the mean velocity gradients; and the Yakhot–Orszag RNG (renormalization group) [262]-based model of Rubinstein and Barton [263] who apply the renormalization group to derive a non-linear algebraic Reynolds stress model of anisotropic turbulence in which the Reynolds stresses are again quadratic functions of the mean velocity gradients. The model results from a perturbation expansion that is truncated systematically at second order with subsequent terms contributing no further information. The resulting turbulence model,

$$\tau_{ij} = -\frac{2}{3}k\delta_{ij} + \nu_{\mathrm{T}}\left(\frac{\partial\overline{u}_i}{\partial x_j} + \frac{\partial\overline{u}_j}{\partial x_i}\right) - \frac{k}{\varepsilon}\nu_{\mathrm{T}}\left[0.4\left(\frac{\partial\overline{u}_i}{\partial x_m}\frac{\partial\overline{u}_j}{\partial x_m}\right)^* + 1.24\left(\frac{\partial\overline{u}_i}{\partial x_m}\frac{\partial\overline{u}_m}{\partial x_j} + \frac{\partial\overline{u}_j}{\partial x_m}\frac{\partial\overline{u}_m}{\partial x_i}\right)^* - 0.16\left(\frac{\partial\overline{u}_m}{\partial x_i}\frac{\partial\overline{u}_m}{\partial x_j}\right)^*\right]$$

applies to both low and high-Reynolds-number flows without requiring wall functions or ad hoc modifications of the equations. All constants are derived from the renormalization group procedure; no adjustable constants arise. k is the turbulence kinetic energy, $\overline{u}_i$ is the mean velocity, and ν_T is the eddy viscosity. (*) indicates the deviatoric part, so, for instance,

$$\left(\frac{\partial \overline{u}_i}{\partial x_m}\frac{\partial \overline{u}_j}{\partial x_m}\right)^* = \frac{\partial \overline{u}_i}{\partial x_m}\frac{\partial \overline{u}_j}{\partial x_m} - \frac{1}{3}\delta_{ij}\frac{\partial \overline{u}_n}{\partial x_m}\frac{\partial \overline{u}_n}{\partial x_m}$$

The model permits inequality of the Reynolds normal stresses, a necessary condition for calculating turbulence-driven secondary flows in non-circular ducts. Their ideas are anchored to a large extent on the original thinking of Yakhot and Orszag [262] who developed the dynamic renormalization group (RNG) method for the study of hydrodynamic turbulence. The RNG theory is quite remarkable in that it does not include any experimentally adjustable parameters, as was the case for the Rubinstein and Barton [263] theory based on RNG, and yields numerical values for important constants of turbulent flows such as the Kolmogorov constant, the turbulent Prandtl number for high-Reynolds-number heat transfer, and the von Karman constant, which come out of the analysis without any recourse to experiments. RNG procedure uses dynamic scaling and invariance together with iterated perturbation methods and allows the evaluation of transport coefficients and transport equations for the large-scale (slow) turbulence modes. Yakhot and Orszag [262] derived a differential transport equation on this basis, which they claim to be particularly useful near walls.

In the non-linear k-ε model of Speziale, the Reynolds stress τ_{ij} is defined by

$$\tau_{ij} = \frac{2}{3}k\delta_{ij} - 2C_\mu \frac{K^2}{\varepsilon}\overline{D}_{ij} - 4C_D C_\mu^2 \frac{K^3}{\varepsilon^2}\left(\overset{\nabla}{\overline{D}}_{ij} + \overline{D}_{ik}\overline{D}_{kj} + \frac{1}{3}\overline{D}_{kl}\overline{D}_{kl}\delta_{ij}\right)$$

where $\overset{\nabla}{\overline{D}}_{ij}$ represents the frame indifferent *Oldroyd derivative* of the mean rate of strain tensor $\overline{D}_{ij}$ and ε is the turbulent dissipation rate.

$$\overset{\nabla}{\overline{D}}_{ij} = \frac{\partial \overline{D}_{ij}}{\partial t} + \overline{u}_k \frac{\partial \overline{D}_{ij}}{\partial x_k} - \frac{\partial \overline{u}_i}{\partial x_k}\overline{D}_{kj} - \frac{\partial \overline{u}_j}{\partial x_k}\overline{D}_{ki}$$

In fully developed duct flow, the axial mean velocity with this model gives rise to a non-zero normal Reynolds stress difference $(\tau_{yy} - \tau_{xx}) \neq 0$ thus enabling the prediction of secondary flows. $C_\mu = 0.09$ and $C_D = 1.68$ are empirical constants. The eddy-viscosity relation of the standard k-ε model is recovered in the limit as $C_D \to 0$.

6.1.2 General Classification and Analysis of Turbulent Secondary Flows

In general, secondary flows can be generated by a combination of the effects of normal Reynolds stress differences, streamline curvature, and body forces arising from the system rotation. Thus, a general classification of turbulent secondary flows may be listed as follows: (1) turbulence-driven secondary flows in *straight ducts* of *non-circular cross section*, (2) turbulent secondary flows in *curved circular* pipes, (3) turbulent secondary flows in *curved ducts* of *non-circular cross section*, and (4) turbulent secondary flows in *rotating ducts* of *non-circular cross section*.

Two-equation turbulence models with an anisotropic eddy viscosity yield acceptable predictions for fully developed secondary flows in straight rectangular ducts. Speziale et al. [264] demonstrate this by numerical comparisons with Reynolds stress closure models. However, for *developing* turbulent secondary flows in *straight* non-circular ducts, where history effects are important, a full Reynolds stress closure is needed for a better and more complete description. The prediction of *curved not fully developed* turbulent pipe flows where both history and near-wall effects play a role, like in the case of a circular pipe U-bend, a second-order closure with a near-wall turbulence model is required for acceptable modeling such as the Reynolds stress closure model of Launder et al. [255] (see Appendix II) with the near-wall turbulence model of Lai and So [265] (see Appendix II). With all second-order closure models, closure is based on the Reynolds stress transport equations, which are derived rigorously from the Navier–Stokes equations, Hinze [250]. These equations read in a Cartesian frame for a non-rotating duct,

$$\frac{\mathrm{D}\tau_{ij}}{\mathrm{D}t}=\frac{\partial\tau_{ij}}{\partial t}+\overline{u}_k\frac{\partial\tau_{ij}}{\partial x_k}=-\tau_{ik}\frac{\partial\overline{u}_j}{\partial x_k}-\tau_{jk}\frac{\partial\overline{u}_i}{\partial x_k}+\frac{\partial C_{ijk}}{\partial x_k}-\Pi_{ij}+\varepsilon_{ij}+\nu\,\Delta\tau_{ij} \tag{6.8}$$

$$C_{ijk}=\overline{u_i'u_j'u_k'}+\overline{p'u_i'}\delta_{jk}+\overline{p'u_j'}\delta_{ik}$$

$$\Pi_{ij}=\overline{2p'D\left(u_i',u_j'\right)}=\overline{p'\left(\frac{\partial u_i'}{\partial x_j}+\frac{\partial u_j'}{\partial x_i}\right)}$$

$$\varepsilon_{ij}=2\nu\overline{\frac{\partial u_i'}{\partial x_m}\frac{\partial u_j'}{\partial x_m}}$$

$$\mathcal{D}_{ij}^T=\frac{\partial\left(\overline{u_i'u_j'u_k'}\right)}{\partial x_k}$$

$$\mathcal{P}_{ij}=\frac{\partial\left(\overline{p'u_i'}\delta_{jk}+\overline{p'u_j'}\delta_{ik}\right)}{\partial x_k}$$

where $\overline{u}_i$, $\overline{u_i'}$, and $\overline{p'}$ represent the mean and fluctuating velocities and fluctuating pressure, respectively. The higher-order turbulence correlations $\mathcal{D}_{ij}^T,\mathcal{P}_{ij},\Pi_{ij},\varepsilon_{ij}$ that appear on the r.h.s. of Eq. (6.8) are called the third-order diffusion correlation, the pressure–diffusion correlation, the pressure–strain correlation (or the pressure–redistribution correlation), and the dissipation rate correlation, respectively. Closure is obtained by tying them to the mean velocity gradients $\nabla\mathbf{u}$, the Reynolds stresses $\boldsymbol{\tau}$, the spatial gradients $\nabla\boldsymbol{\tau}$ of the Reynolds stresses, and the length scale of turbulence through turbulence closure models. Some of these models are reviewed in Appendix II for completeness. A review of the analytical methods for the development of Reynolds stress closures is given by Speziale [266].

For homogeneous turbulent flows, where the mean velocity gradients are spatially uniform, at high Reynolds numbers, the dissipation is approximately isotropic and the Reynolds transport equation (Eq. 6.8) reduces to

$$\frac{\mathrm{D}\tau_{ij}}{\mathrm{D}t} = \frac{\partial \tau_{ij}}{\partial t} + \bar{u}_k \frac{\partial \tau_{ij}}{\partial x_k} = -\tau_{ik}\frac{\partial \bar{u}_j}{\partial x_k} - \tau_{jk}\frac{\partial \bar{u}_i}{\partial x_k} - \Pi_{ij} + \frac{2}{3}\varepsilon\delta_{ij} \tag{6.9}$$

$$\varepsilon = \nu\overline{\frac{\partial u_i'}{\partial x_m}\frac{\partial u_i'}{\partial x_m}}$$

where ε is the scalar dissipation rate. In both near-wall turbulence and homogeneous turbulence away from the walls, Π_{ij} contains directional information and plays a pivotal role in the evolution of the Reynolds stress tensor. As it is the only unknown correlation containing directional information in homogeneous flows, researchers relied on homogeneous turbulence for the testing and calibration of pressure–strain models. In wall-bounded turbulence, Reynolds stress closures based on Eq. (6.9) fall short of modeling turbulence statistics. The reason lies in the structure of Eq. (6.9), which really amounts to the neglect of the pressure–diffusion term $\mathcal{P}_{ij}$ on the grounds that for high-Reynolds-number flows $|\mathcal{P}_{ij}| \ll |\mathcal{D}_{ij}^T|$. Furthermore, since viscosity does not appear explicitly in the Poisson equation for the fluctuating pressure p',

$$\Delta p' = -2\frac{\partial u_j}{\partial x_i}\frac{\partial u_i}{\partial x_j} - \frac{\partial u_i}{\partial x_j}\frac{\partial u_j}{\partial x_i} + \overline{\frac{\partial u_i'}{\partial x_j}\frac{\partial u_j'}{\partial x_i}}$$

it was mistakenly argued that the high-Reynolds-number form of the redistribution model is also applicable for near-wall flow calculations. As a result, only the high-Reynolds-number isotropic form of the viscous dissipation rate model of Kolmogorov [267] is modified to account for anisotropic behavior near a wall. Most models for Π_{ij} built on this premise perform well away from the wall. However, neglecting $\mathcal{P}_{ij}$ means that the tensor $(\mathcal{P}_{ij} - \Pi_{ij})$ with a non-zero trace is now replaced with Π_{ij} a tensor with zero trace, which requires that the models for Π_{ij} are traceless as well leading to unacceptable finite near-wall values for the two components of the redistribution tensor Π_{ij}. That in turn severely limits the ability of the Reynolds stress closure models based on Eq. (6.9) to model the turbulence statistics near the wall successfully. Lai and So [255] suggest that a model for $(\mathcal{P}_{ij} - \Pi_{ij})$ should be able to balance the deficit in the term $(\nu\Delta\tau_{ij} - \varepsilon_{ij})$ near the wall but should approach asymptotically Π_{ij} away from the wall. They show that $\frac{\mathrm{D}\tau_{ij}}{\mathrm{D}t}$, $\mathcal{D}_{ij}^T$, and $\left(\tau_{jm}\frac{\partial \bar{u}_i}{\partial x_m} + \tau_{im}\frac{\partial \bar{u}_j}{\partial x_m}\right)$ go to zero at the wall like y^n, $n \geq 3$ where y is the coordinate normal to the wall, and argue that $\nu\Delta\tau_{ij}$ and ε_{ij} are dominant near the wall. As $\frac{D\tau_{ij}}{Dt}$ goes to zero at the wall, the difference in $(\nu\Delta\tau_{ij} - \varepsilon_{ij})$ is to be compensated by the remaining terms $\left(\mathcal{P}_{ij} - \Pi_{ij}\right) = \left(\overline{u_i' \frac{\partial p'}{\partial x_j}} + \overline{u_j' \frac{\partial p'}{\partial x_i}}\right)$. Based on this understanding, Lai and So [265] propose a methodology to derive an asymptotically correct model for the velocity–pressure–gradient correlation and derive a model that has the property of approaching the high-Reynolds-number model for pressure redistribution far away from the wall. A similar analysis is carried out on the viscous dissipation term and asymptotically correct near-wall modifications are proposed. Lai and So [265] obtain excellent agreement with measured and simulated near-wall turbulence statistics especially the anisotropic behavior of the normal stresses. They point out that the dissipation rate ε_{ij} has a significant effect on the calculated Reynolds stresses and that the present estimates can be improved upon by using direct simulation (DNS) data as it becomes available to help formulate a more realistic dissipation rate equation and a near-wall closure for the Reynolds stress equations using the approach they initiated. A full description of the Lai and So [265] model is given in Appendix II.

6.2 Secondary Flows of Non-Brownian Suspensions

There are many applications in industry where the processing and transport of suspensions is an important operation. Suspension types are quite diverse and include cement, paint, printing inks, ceramics, reinforced polymer composites, coal slurries, drilling muds, liquid abrasive cleaners, foodstuffs, medicinal liquids, and suspensions with deformable particles such as emulsions and blood. Applications in the area of energy include hydraulic fracturing technology and processing of solid rocket propellants. Suspensions are frequently encountered in natural phenomena as well such as mud flows and landslides. Suspensions present diverse rheological behavior such as shear thinning, shear thickening, and thixotropy depending on the microstructure of the materials, the nature of the interactions involved, and the properties of the carrier liquid. Engineering processes often require that a flowing suspension be supplied at a specified location with a prescribed particle concentration. Rheology of suspensions remains an active research field, and no unifying view has been proposed yet.

There are two types of secondary flows with suspended materials, those generated by normal stress effects in laminar flow of suspensions in tubes of cross section other than circular and the class of secondary flows in turbulent flow of suspensions generated either by body forces or by the anisotropy of the Reynolds stress tensor. The latter is equally likely to happen in turbulent flow in tubes of circular and non-circular cross section depending on the shape of the cross section, the roughness distribution on the boundary, and the suspended particle distribution within the cross section. Both of these classes of secondary flows have relevance to industrial operations and will be discussed in Sects. 6.2.6, 6.2.7, and 6.2.8. But first, the all-important issues of the shear viscosity and of the normal stress measurements in suspensions will be addressed in the next two Sects. 6.2.1 and 6.2.2 and some suspension models in Sect. 6.2.3.

6.2.1 Shear Viscosity of Non-Brownian Suspensions

Given the difficulties involved in formulating a general approach to suspension rheology, recent research up to this writing focused on neutrally buoyant non-Brownian hard spherical particle suspensions. Traditionally, suspensions have been modeled as isotropic Newtonian fluids with effective viscosities that vary with concentration when calculating the velocity distribution through a conduit, Barnes et al. [268]. Viscosity of suspensions of solid particles in Newtonian liquids for the case of dilute dispersed suspensions of 10 % and less phase volume ϕ (fraction of space occupied by the suspended material) is calculated by the well-known Einstein [269–271] formula dating back to 1906

$$\eta = \eta_s(1 + 2.5\phi)$$

where η and η_s represent the viscosities of the suspension and of the suspending medium, respectively. Phase volume is an important concept because the rheology depends to a great extent on the hydrodynamic forces which act on the surface of particles or aggregates of particles irrespective of the particle density. Einstein's equation neglects the effect of particle interaction and does not account for the effects of particle size and of particle position. The interaction between neighboring particles is accounted for by higher-order terms in ϕ. However, the only tractable theory is for extensional flow because only in this type of flow can the relative position of particles be accounted for analytically, Batchelor and Green [272, 273] and Batchelor [274]

$$\eta = \eta_s\left(1 + 2.5\phi + 6.2\phi^2\right) \tag{6.10}$$

where the viscosities must now be interpreted as extensional viscosities.

It should be noted that dilute suspensions are of little industrial relevance; however, the work is important in laying the foundations of viscosity computations for concentrated suspensions. Dilute suspension theory covers the range below 10 % phase volume, and this accounts for no more than 40 % increase in viscosity over the continuous phase. Current understanding of non-colloidal dilute suspension rheology is based on the work of Batchelor and Green [272, 273] and Batchelor [274] whose work collaboratively led to the derivation of Eq. (6.10), the viscosity of a dilute suspension to order $O(\phi^2)$.

The concentrated or dense suspensions are defined as the regime where the average separation distance between particles is smaller than the particle size. In this regime multiple-body and short-range interactions give rise to strong non-Newtonian effects. Dense suspensions of neutrally buoyant non-Brownian hard spheres are found to exhibit a Newtonian relation between the stress and shear rate, thus allowing the definition of a relative suspension viscosity function of the particle concentration. This was first suggested by the dimensional analysis of Krieger [275, 276] and later confirmed by both experimental observations and numerical simulations. As may be expected intuitively, suspension viscosity increases with the concentration, and in the dense regime, a divergence of viscosity is observed at a maximum packing fraction ϕ_m. The maximum packing fraction is defined as the maximum solid volume fraction for which the suspension exhibits fluid behavior. This behavior can be captured by empirical functions kept consistent with the viscosity predictions for dilute suspensions. The value of ϕ_m ranges over $0.58 < \phi_m < 0.68$ due to the large scatter in the data nor is there any agreement in the research community on the form of the viscosity curve close to the divergence.

At very low Reynolds numbers in simple shear flow, a linear relation exists between the shear stress $\tau = \Sigma_{xy}$ and the shear rate $\dot{\gamma}$ with a quasi-Newtonian effective suspension viscosity $\eta(\phi) = \frac{\Sigma_{xy}}{\dot{\gamma}}$, $\eta(\phi) = \eta_f \eta_s$. Note that it is customary in suspension research to designate the total stress (fluid + particle) by Σ. η_f is the viscosity of the suspending medium and η_s (sometimes called relative viscosity) is a function of the volume fraction ϕ which diverges when approaching the maximum packing fraction ϕ_m, Stickel and Powell [277]. Notably Leighton and Acrivos [278] and Krieger [275, 276] proposed empirical expressions for η_s. The relationship derived by Krieger [275, 276] through dimensional considerations is still widely in use and reads as

$$\eta_s = \left(1 - \frac{\phi}{\phi_m}\right)^{-1.82} \tag{6.11}$$

There are other correlations in the literature developed by various authors. In particular the Maron–Pierce empirical equation [with the exponent in Eq. (6.11) replaced by (−2)]; the Eiler correlation,

$$\eta_s = \left[1 + 1.5\left(\frac{\phi\phi_m}{1 - \phi}\right)\right]^2;$$

the Morris and Bouley [279] correlation,

$$\eta_s = 1 + 2.5\phi_m\left(1 - \frac{\phi}{\phi_m}\right)^{-1} + K_s\left(\frac{\phi}{\phi_m}\right)^2\left(1 - \frac{\phi}{\phi_m}\right)^{-2}; \tag{6.12}$$

where K_s is a curve fitting coefficient; and the Zarraga et al. [280] correlation,

$$\eta_s = e^{-2.34\phi}\left(1 - \frac{\phi}{\phi_m}\right)^{-3}$$

are in frequent use in data matching.

6.2.2 Recent Normal Stress Measurements

The role of normal stresses in suspension rheology has been much less documented even though, as is the case with viscoelastic fluids, normal stress differences in suspensions are of fundamental importance in many situations since they can induce secondary flows (see Sects. 6.2.6, 6.2.7, and 6.2.8) and instabilities. They are also closely related to flow-induced microstructure and to particle migration. The origin of normal stress differences is far from being understood completely. It is commonly believed that they arise from hydrodynamic interactions or direct contact between particles in the case of non-Brownian (non-colloidal) particles. Microstructure changes under shear are the fundamental mechanism behind the development of the observed normal stresses in suspensions. A crude understanding of the physics behind nascent normal stresses can be gained by considering the interaction between a pair of particles in Stokes flow, Fig. 6.1. This line of thinking dates back to Parsi and Gadala-Maria [281].

Stokes flow around a single *smooth* sphere is perfectly symmetric and reversible, Fig. 6.1a. Particles are kept apart by lubrication and the compressive and extensional parts of the flow, aft and fore respectively, cancel out. However, if the spheres are *rough* and the minimum approach distance for the pair is smaller than the roughness, the spheres come into contact in the compressional region of the flow and move apart in the extensional region creating asymmetry and irreversibility, Fig. 6.1b. Asymmetry leads to weaker hydrodynamic interactions in the extensional

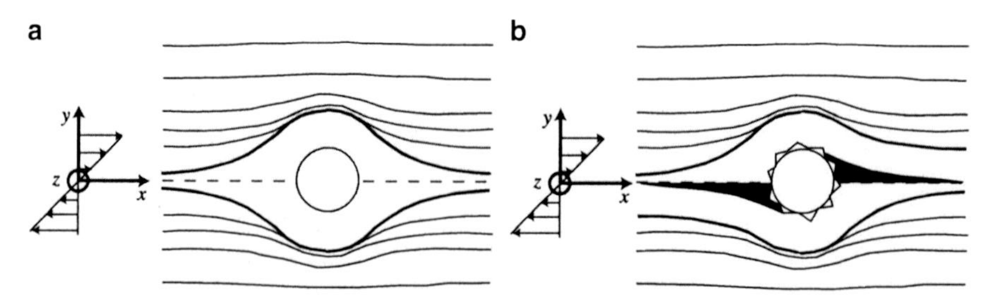

Fig. 6.1 Particle trajectories around a sphere in a linear shear flow: (**a**) symmetry for perfectly smooth spheres and (**b**) asymmetry for rough spheres with depletion in the extensional quadrants indicated in *black*. The limiting (open) trajectories are indicated by *thick lines* (reprinted from Coutourier et al. [287] with permission)

region than in the compressional region due to particle depletion shown in black in Fig. 6.1b. This explains conceptually the development of normal stresses. The negative first normal stress difference $N_1 = \Sigma_{xx} - \Sigma_{yy}$ is attributed to the deficit in hydrodynamic interactions at the rear of the collision that is in the rear extensional region. The absence of shear in the planes parallel to Oxz and additional stresses in the Oxy plane created by particle interactions lead to a negative second normal stress difference $N_2 = \Sigma_{yy} - \Sigma_{zz}$. It is predicted that in dilute regime, the normal stress coefficients α_1 and α_2 vary as ϕ^2 and depend on the surface roughness, Brady and Morris [282], Wilson [283]. In the dense regime, Sierou and Brady [284, 285] have shown through Stokesian dynamics simulations that the magnitude of normal stress differences depends to a large extent on frictional particle contact. It is generally accepted that for suspensions of non-Brownian hard spheres in the Stokes regime, N_1 and N_2 are proportional to the shear stress $\tau = \Sigma_{xy}$. The normal stress coefficients α_1 and α_2 are functions of ϕ only and are defined as $\alpha_i(\phi) = {}^{N_i}/_{\Sigma_{xy}}, \quad i = 1, 2$, where N_i are the normal stresses and Σ_{xy} represents the shear stress.

Suspensions of rigid non-colloidal particles are known to exhibit a strong negative second normal stress difference, in particular when the suspension is concentrated, giving rise to secondary currents. Recent careful measurements reveal that normal stress effects strongly manifest themselves only above a volume fraction of approximately 22 %, Boyer et al. [286]. They have shown that a transition seems to occur at $\phi \approx 0.2$ with a strong linear increase of $\left(\alpha_2 + \frac{\alpha_1}{2}\right)$ for $\phi \geq 0.2$. The mechanics of neutrally buoyant suspensions of non-Brownian hard spherical particles in a Newtonian liquid is shaped only by hydrodynamic and contact forces. That the normal stress effects start manifesting themselves strongly around a volume fraction of 0.22 gives support to the theory that the behavioral change can be attributed to contact forces. The concentrated or dense suspensions are defined as the regime where the average separation distance between particles is smaller than the particle size, Stickel and Powell [277]. The suspension microstructure in this regime is poorly understood, and constitutive equations relating stress to the strain rate are not generally known.

There are only a few reported measurements of normal stresses in dense suspensions. Until the very recent attempts by Boyer et al. [286], Coutourier et al. [287], and Dbouk et al. [288], there were only two measurements of normal stresses in dense suspensions available in the literature, Zarraga et al. [280] and Singh and Nott [289], done very early in this century. Zarraga et al. [280] were the first to use a rotating rod immersed in a bath of suspension to study normal stress differences. They used glass spheres in their experiments. By combining measurements of the suspension normal stress coefficient combination $\left(\alpha_2 + \frac{\alpha_1}{2}\right)$ obtained from free surface profile matching in rod-dipping experiments with measurements of the net thrust force F directly proportional to $(N_1 - N_2)$ on one of the disks of radius R in the torsional parallel plate,

$$N_1 - N_2 = \frac{2F}{\pi R^2}\left(1 + \frac{1}{2}\frac{d\ln F}{d\ln \dot{\gamma}_R}\right)$$

they were able to obtain numerical values for α_1 and α_2. They have shown that the normal stress differences N_1 and N_2 are both negative and proportional to the shear stress or the shear rate with $|N_1| < |N_2|$. They propose the following correlation:

$$\frac{1}{\eta_s}\left(\alpha_2 + \frac{\alpha_1}{2}\right) = 1.16\phi^3 e^{2.34\phi}$$

Their results have been qualitatively confirmed by Singh and Nott [289] who used data from cylindrical Couette and parallel-plate rheometers with polymethyl methacrylate (PMMA) spheres suspended in a Newtonian medium. However, the parallel-plate rheometer is unsuitable for suspension rheometry as holding a suspension in sufficient quantity between the plates requires a wide enough gap, which often promotes drainage at the boundary. In their experiments, Singh and Nott [289] immersed the rotating disks of the parallel-plate rheometer in a bath of suspension to prevent drainage. The direct data obtained from the experiments of Zarraga et al. [280] and Singh and Nott [289] are different. The former obtain α_1 and α_2 whereas the latter's combination of the Couette and parallel-plate rheometer experiments yields N_1 and N_2. Singh and Nott [289] find that both N_1 and N_2 are negative and are increasing functions of the volume fraction with N_1 three times smaller than N_2. Although there is a qualitative agreement of their respective data quantitatively, there is a difference of a factor of 2 between the results of Zarraga et al. [280] and Singh and Nott [289], which clearly spells out that the matter is far from settled. The methods used by these authors may introduce significant errors as they require at least a set of data from two rheometric experiments each of which yields linear combinations of the normal stress coefficients α_1 and α_2. Coutourier et al. [287] circumvented this problem by using the tilted trough rheometer to obtain accurate measurements of the second normal stress coefficient α_2.

The classical *rod-climbing* problem *on a viscoelastic fluid* has a long history starting with the independent efforts of Ericksen and Serrin to seek a solution but was put on solid grounds only by Joseph and Fosdick [290], Joseph et al. [291], and Beavers and Joseph [292]. The shape of the free surface on a liquid filling the semi-infinite space between two concentric cylinders which rotate at different speeds is a striking manifestation of the *normal stress effect*. The shearing of the liquid induces *normal stresses* along and perpendicular to cylinder generators on planes which are perpendicular to the planes of primary shear. These normal stresses are larger than the primary shear, and in the regions of greatest shear, the free surface is pushed up along the cylinder generators. The analysis due in principle to Ericksen for a general simple fluid, which only determines the direction of the climbing in a qualitative sense, is reported by Truesdell and Noll [115]. Serrin [293] studied this problem assuming the fluid obeys the Reiner–Rivlin CE. His analysis has obvious shortcomings in hindsight because of the simplifying assumptions he made such as the neglect of the unbalanced shear stresses at the free surface and replacing the two material functions of the Reiner–Rivlin theory with constants. Joseph and Fosdick [290] develop a systematic construction in series for the shape of the free surface on a simple fluid as well as for the induced secondary motion from a perturbation of a state of rest with perturbation parameter the angular velocity of one of the cylinders. The analysis is based on the domain perturbation method developed earlier by Joseph [294] and involves solving a boundary value problem in a given region of space by mapping it onto a region of much simpler shape. The mapped problem is then expanded in a power series in the parameter characterizing the domain deformation. The perturbation problems that arise in the expansion are solved successively in the simpler standard region, and the resulting series is mapped back into the original domain. Beavers and Joseph [292] report on the development of practical methods of viscometry to characterize non-Newtonian fluids in slow flow. They show that measurements of the free surface near rods rotating in STP and polyacrylamide solutions are accurate, reproducible, and in excellent agreement with a theory of rod-climbing developed earlier, Joseph and Fosdick [290]. Results are presented that establish the theory and experiment as a viscometer for determining the values of Rivlin–Ericksen constants that arise in the theory of slow flow. The variation of these constants with temperature in the sample of STP used in the experiments is explicitly and accurately determined. The experiments in STP show that there is a range of rotational speeds for which the behavior of STP may be well described by the CE characterizing *the fluids of grade four*. In later years, the rod-climbing phenomenon (or the Weissenberg effect) in an eccentric cylinder geometry was investigated by Siginer [295] theoretically and by Siginer and Beavers [296] experimentally, and Siginer [297, 298] advocated the use of the eccentric cylinder geometry as a free surface rheometer.

Following the lead of Zarraga et al. [280], the classical rod-climbing experiment (rotating-rod rheometry) was used by Boyer et al. [286] to characterize and measure the normal stresses in a suspension of non-colloidal neutrally buoyant (absence of sedimentation) rigid spheres dispersed in a Newtonian fluid. It should be noted that "rod-climbing" is a misnomer in this case as *normal stresses* cause the free surface on the suspension to sink near the rod rather than forcing it to climb up the rod as would be the case with polymeric fluids, that is, the negative values of normal stress differences in concentrated suspensions give rise to a negative surface deflection or rod-dipping effect. Boyer et al. [286] used an experimental set-up adapted from the rotating-rod rheometer introduced by Beavers and Joseph [292] and employed a sophisticated optical profilometry method to measure the free surface deflections. They also observed a phenomenon not reported by Zarraga et al. [280], the drifting in time of the free surface measurements, a time-dependent phenomenon typical of non-Brownian suspensions. Specifically the dip next to the rotating rod is gradually filled in time due to particles slowly migrating from regions of high shear rate to regions of low shear rate, that is, away from the rotating rod. The predictions of the migration based on a suspension balance model in which the particle migration flux is related to the *normal stresses* of the suspension agree well with observed changes.

The second normal stress difference in suspensions of non-Brownian neutrally buoyant rigid spheres dispersed in a Newtonian fluid was measured by Coutourier et al. [287]. They got their inspiration from the studies of Wineman and Pipkin [299], Tanner [300], and Sturges and Joseph [301] who investigated the normal stress effects with viscoelastic fluids in a similar experimental set-up. The approach is based on the determination of the shape of the free surface on the suspension flowing in a tilted trough. The free surface on a Newtonian fluid is flat. But with a suspension, the normal stress effect produces a distortion of the plane free surface of the Newtonian flow. Wineman and Pipkin were the first to suggest that the free surface on the viscoelastic liquid flowing in a trough could be used to obtain information on the constitutive constants of the fluid. They show that the free surface on a viscoelastic fluid will bulge out if $2\alpha_1 + \alpha_2 < 0$, where α_1 and α_2 are the first and second Rivlin–Ericksen constants (note that the normal stress coefficients α_1 and α_2 for viscoelastic fluids and suspensions are different in definition). This combination of parameters gives the limiting value of the second normal stress difference for viscoelastic fluids for the shear rate $\dot{\gamma}$ tending to zero,

$$\widetilde{N}_2 = 2\alpha_1 + \alpha_2 = \lim_{\dot{\gamma} \to 0} \frac{N_2\left(\dot{\gamma}\right)}{\dot{\gamma}}$$

This bulging observed with all viscoelastic fluids implies that $N_2\left(\dot{\gamma}\right)$ is negative when the shear rate is small. Sturges and Joseph [301] attempt to formulate a theory of viscometry for slow steady motions which comes down to finding guidance for experiments through analysis in the form of relationships for Rivlin–Ericksen constants that appear in the expressions for the extra-stress in slow steady motion. They calculate the solution in terms of infinite series of the powers of the tilt angle β and show that secondary motions do not appear until $O(\beta^6)$ when the trough is infinitely deep or has a semi-circular shape. In these two cases, simple formulas are given relating the shape of the free surface to the Rivlin–Ericksen constants of the first-, second-, third-, and fourth-order fluids (see Sect. 3.3.2 in this book and section 2.4.5 in Siginer [73]). A free surface rheometer of this type was also proposed by Siginer [302] in the early 1990s to guide experiments. He studied the free surface deformation on a viscoelastic fluid in a vat with steadily rotating bottom and derived an expression relating combinations of the Rivlin–Ericksen constants $\alpha_1 + \alpha_2$ and $(2\alpha_1 + \alpha_2)$ to the free surface distortion at the lowest significant order (second order) by the method of domain perturbation. The surface is shaped by the pressure field set-up by competing centrifugal forces and normal stresses. It turns out that the combination $\alpha_1 + \alpha_2$ is far more influential in shaping the free surface than the second normal stress difference N_2 which corresponds to $(2\alpha_1 + \alpha_2)$. Matching experimental and predicted free surface shapes yields the values of the Rivlin–Ericksen constants. The values of α_1 and α_2 thus obtained can be checked against the values determined either by cone and plate or parallel-plate rheometry both of which yield $(N_1 - N_2)$ and N_2 through torque and normal thrust measurements. The normal stress functions for viscoelastic fluids are given by the well-known expressions, Truesdell and Noll [115],

$$N_1\left(\dot{\gamma}\right) = \alpha_2\left(\dot{\gamma}\right)^2 + \boldsymbol{O}\left[\left(\dot{\gamma}\right)^4\right]$$

$$N_2\left(\dot{\gamma}\right) = (2\alpha_1 + \alpha_2)\left(\dot{\gamma}\right)^2 + \boldsymbol{O}\left[\left(\dot{\gamma}\right)^4\right].$$

The Rivlin–Ericksen constants are obtained from

$$\lim_{\dot{\gamma}\to 0} \frac{N_1\left(\dot{\gamma}\right)}{\dot{\gamma}} = \alpha_2$$

$$\lim_{\dot{\gamma}\to 0} \frac{N_1\left(\dot{\gamma}\right) - N_2\left(\dot{\gamma}\right)}{\dot{\gamma}} = -2\alpha_1.$$

Siginer and Knight [303] also investigated the swirling free surface flow in cylindrical containers and determined the free surface on a linear fluid in cylindrical containers of variable aspect ratio by domain perturbations and expanding the solution for the meridional field into complex biorthogonal series of the Papkovich–Fadle type.

Very much along the same lines of analysis and thinking, Coutourier et al. [287] use free surface profile matching to determine the second normal stress difference for dense suspensions. The free surface on polymeric fluids flowing down a tilted trough displays a convex parabolic bulge. However, the free surface shape on a non-Brownian suspension flowing in the same trough is very different as the normal stress differences are linear in the shear rate $\dot{\gamma}$ for a non-Brownian suspension, whereas they are quadratic in $\dot{\gamma}$ for polymeric liquids. They deduce the shape of the free surface on a Stokesian suspension of non-Brownian neutrally buoyant spheres characterized by the Péclet number (ratio of viscous forces to Brownian forces) $Pe \to \infty$ and the particle-based Reynolds number $Re_p \to 0$,

$$Pe = \left(\frac{6\pi\eta_f a^3 \dot{\gamma}}{kT}\right) \to \infty, \qquad Re_p = \frac{\rho_p a^2 \dot{\gamma}}{\eta_f} \to 0 \tag{6.13}$$

based on the linearity of the relationship between the total suspension shear stress Σ_{xy} and the shear rate and the linear dependence of the second normal stress difference on the shear stress,

$$\Sigma_{xy} = \eta_s(\phi)\,\dot{\gamma},$$

$$N_2 = \Sigma_{yy} - \Sigma_{zz} = \alpha_2(\phi)\Sigma_{xy}$$

In these, η_f, ρ_p, and a are the suspending fluid viscosity, the particle density, and the particle radius. The particle volume fraction ϕ is determined by $\phi = 4\pi a^3 n/3$ where n is the particle number density. The restriction (Eq. 6.13)$_2$ on the particle Reynolds number $Re_p \ll 1$ based on the suspended particle size is not an overly severe assumption because there are many situations in suspension processing which very closely satisfies this restriction.

The experimental component of the research was conducted with a Newtonian suspending fluid with density closely matched to that of the particles $\rho_f = 1.051$ g cm^{-3} at 25 °C and large viscosity $\eta_f = 2.15$ Pa s. Two batches of polystyrene spheres of the same density $\rho_p = 1.049 \pm 0.003$ g cm^{-3} and a typical surface roughness height of 0.2 μm, but two different radii were used. Sedimentation could safely be neglected under these circumstances as the particle settling velocity (Stokes velocity) was in the range of (0.001 mm h^{-1}) and the time scale of the experiments was 10 min. Migration of particles under shear was not a concern either even though it is well known that particles migrate under heterogeneous shear. Coutourier et al. [287] calculate based on the findings of Leighton and Acrivos [278] and Morris and Bouley [279] that strain rates that may lead to migration are of the order of $(W/a)^2 \sim 10^4$, whereas in the experiments, the largest strain is of the order 10. The data coming from the two batches of particles collapse on the same line, thus showing no dependence on the particle size, except perhaps at the highest value of ϕ investigated $\phi = 0.5$ where $(-\alpha_2)$ is found to be smaller for the suspension with the larger spheres. They find that the second normal stress difference for dense suspensions is negative and linear in the shear stress and the ratio of the second normal stress difference to the shear stress increases with increasing volume fraction exhibiting a strong, approximately linear, growth in magnitude with volume fractions ϕ above a volume fraction of $\phi_c = 0.22$. For $\phi < 0.25$, the increase in $(-\alpha_2)$ is slow, whereas for $\phi > 0.25$, there is a strong linear increase with ϕ represented by the correlation 1.4 $(\phi - \phi_c)$. The limit of resolution of the measurement (typically 1 Pa) does not permit detection of normal stresses for $\phi < 0.17$. The data is of the same order of magnitude for $\phi < 0.25$ as the numerical predictions of Sierou and Brady [284] and the analytical predictions of Wilson [283] assuming the roughness height is 0.2 μm the same as the experimentally measured value. There is good agreement as well for $\phi > 0.25$ with the numerical simulations of Sierou and Brady [284] when the frictional effects are taken into account with a coefficient of friction of 0.5. The magnitude of the ratio of the first normal stress difference to the shear stress is found to be very small. They conclude that it is hard to tell if α_1 is positive or negative as the experimental values fluctuate around zero $-0.06 < \alpha_1 < 0.06$; thus N_1 could be either positive or negative but in any case it is $N_1 < 0.06\ \Sigma_{12}$.

An experimental approach to measure both normal stress differences and the particle phase contribution to the normal stresses in suspensions of non-Brownian hard spheres was developed by Dbouk et al. [288] who proposed to determine the normal stress differences from measurements of the radial profile of the second normal stress Σ_{22} (along the velocity gradient direction) in a suspension sheared between two rotating parallel plates. The measurements are carried out for a wide range of particle volume fractions (between 0.2 and 0.5). N_2 is measured to be negative but N_1 is found to be positive. Dbouk et al. [288] measured the pore pressure in a suspension sheared in a cylindrical Couette cell following the more recent and the earlier work of Deboeuf et al. [304] and Prasad and Kytomaa [305], respectively. Thus, the contribution of the particles Σ^p_{22} to the normal stress can be obtained by subtracting the pore pressure P^f from the total normal stress Σ_{22}. In particular, the measurements of Dbouk et al. [288] reveal for the first time that the particle normal stresses are of the same order of magnitude as the second normal stress difference, and they confirm that the magnitude of the particle stress tensor components and their dependence on the particle volume fraction used in the suspension balance model proposed by Morris and Boulay [279] are realistic.

Stokesian dynamics simulations conducted either at large finite Pe with no repulsive forces, Phung et al. [306], or at infinite Pe with short-range repulsive forces, Yurkovetsky [307], all point out that both N_1 and N_2 are negative and the second normal stress difference is larger of the two in absolute value $|N_2| > |N_1|$. Yurkovetsky's research, in particular, indicates that all normal stress components are negative, indicating that the stress is compressive in a concentrated $(0.45 < \phi < 0.52)$ near-hard-sphere suspension. Microstructural analysis of Brady and Morris [282] confirms that all normal stress components are negative (compressive) and that $|N_2| > |N_1|$. They state "It is clear that anisotropic compressive normal stresses are present in sheared suspensions, but there remains much uncertainty regarding their precise form." However, these findings were contested recently by Dbouk et al. [288] who determine that $N_1 > 0$.

The latest contribution at the time of this writing to this area of hot topic research has been made by Garland et al. [308] who measured in a cylindrical Taylor–Couette device the shear-induced radial normal stress in a suspension of neutrally buoyant non-Brownian spheres immersed in a Newtonian viscous liquid. They find that the radial normal stress of the particle phase Σ^p_{rr} is independent of the particle size, and its ratio to the suspension shear stress increases quadratically with ϕ, in the range $0 < \phi < 0.4$. Further, its magnitude is about an order of magnitude higher than the second normal stress difference of the suspension in agreement with the theoretical particle pressure predictions of Brady and Morris [282] for small ϕ supporting the argument that the particle phase normal stress Σ^p_{rr} is due to asymmetric pair interactions under dilute conditions, and may not require many-body effects.

6.2.3 *Macroscopic Models for Suspension Flow and Stokesian Dynamics Simulations*

Shear-induced migration of particles in situations where inhomogeneous stress or shear fields are present is common in practical situations such as pressure-driven flow of a suspension in tubes and channels and is a distinctly different phenomenon from the migration of particles due to inertial forces observed by Segré and Silberberg [309–311] who showed that at very low Reynolds numbers of $O(1)$, a rigid sphere transported along in Poiseuille flow through a tube is subject to radial forces which tend to carry it to an equilibrium position at about 0.6 tube radii from the axis, irrespective of the radial position at which the sphere first entered the tube. They further prove that the origin of the forces causing the radial displacements is in the inertia of the moving fluid.

The steady long-time decrease as well as the short-time increase in the effective viscosity of a concentrated suspension in a Couette viscometer was observed for the first time by Gadala-Maria and Acrivos [312], who thus provided the first experimental evidence for the existence of shear-induced migration. Later, Leighton and Acrivos [278] conducted experiments to show that this was due to the migration of particles from the Couette gap into the reservoir at the bottom. Migration of particles in the Couette viscometer is in the direction of the gradient in shear rate perpendicular to the plane of shear. In all other studies on pressure-gradient-driven flow and wide-gap Couette flow, for example, the shear rate gradient is in the same direction as the shear rate. This is unlike narrow gap Couette flow and suspension shear flow between parallel plates where the stress and the rate of strain are uniform, which means that the suspension is independent of position and consequently there is no "preferred" position to be sought by the particles, and any particle moves about randomly as it follows the mean motion. This random motion is due to the chaotic nature of the particle evolution equations in concentrated suspensions and gives rise to a diffusive behavior, the shear-induced self-diffusivity of non-Brownian particles, measured experimentally by Eckstein et al. [313] and Leighton and Acrivos [314]. The former authors determine the self-diffusion coefficients using a concentric-cylinder Couette apparatus for lateral dispersion of spherical and disk-like particles in linear shear flow of a slurry at very low Reynolds number. The method requires the direct measurement of the lateral position of the marked particle. In contrast, Leighton and Acrivos [314] determine the coefficient of shear-induced particle self-diffusion in concentrated suspensions of solid spheres from the measured variations in the time taken by a single marked particle in the suspension to complete successive circuits in a Couette device. Their experiment is simpler than that of Eckstein et al. [313] constrained that it is by wall effects at high particle concentration. Leighton and Acrivos [314] found the diffusion coefficient to be proportional to the product $\dot{\gamma} a^2$, where $\dot{\gamma}$ is the shear rate and a the particle radius. In the dilute limit when the particle concentration $\phi \rightarrow 0$, the diffusion coefficient assumes the asymptotic form $0.5\dot{\gamma} a^2 \phi^2$.

Nott and Brady [315] were concerned with the *net* migration of particles in a suspension exposed to inhomogeneous stress or shear fields. Leighton and Acrivos [278] argued that small-scale surface roughness of the particles leads to irreversible motion during interparticle interaction and that *net* migration arises due to a greater number of interactions on one side of a particle than the other in a flow field with a gradient in shear rate, shear stress, or particle concentration. But Nott and Brady [315] show that *net* irreversible migration is produced even when suspended particles are perfectly smooth hard spheres and that surface roughness is not the sole mechanism leading to *net* irreversible migration. They use Stokesian dynamics simulations introduced by Brady and Bossis [316, 317] to perform dynamic simulations of the pressure-driven flow of a suspension of perfect hard spheres at zero Reynolds number in a rectangular channel. Stokesian dynamics approach allows the computation of the many-body long-range hydrodynamic interactions as well as the short-range lubrication interactions in Stokes flow. Nott and Brady [315] conducted simulations along the same lines as Brady and Bossis for monodisperse non-Brownian hard spheres in the absence of external forces with about 50 particles. The initial particle configurations were generated by a Monte Carlo technique; the interior particles were first placed in a regular array between the walls and then successively given small random displacements until a uniform distribution was achieved. A short-range repulsive force between interior particles was used in all simulations. These forces were chosen to be of the form *previously* used by Durlofsky and Brady [318],

$$\mathbf{F}_{\alpha\beta} = F_0 \tau \frac{e^{\frac{\in}{\tau}}}{1 - e^{\frac{\in}{\tau}}} \mathbf{e}_{\alpha\beta}$$

where $\mathbf{F}_{\alpha\beta}$ is the force exerted on sphere α by sphere β, F_0 is a constant representing the magnitude of the force, $\in$ is the spacing between the surfaces of spheres α and β, $\mathbf{e}_{\alpha\beta}$ is the unit vector connecting the centers of the spheres α and β, and τ is related to the range. The value of τ was in the range $10^2 < \tau < 10^4$ and $\in \sim O(10^{-4})$. The interparticle repulsive force prevents the formation of large clusters by counteracting the strong lubrication forces which tend to keep particles together once they are in close proximity. The purpose of the repulsive force is to qualitatively model the effect of the Brownian motion and surface roughness of particles among others that may counteract the effect of the strong lubrication forces, thus providing a mechanism through which the clusters may be broken. The short-range repulsive force leaves the long-range interactions unaltered. The *net* migration of particles is not affected by the presence or absence of the short-range repulsive forces, and therefore, these forces are not essential for the phenomenon of shear-induced migration. They also derive the time scale, which can be expressed in terms of the channel length L and height H and particle radius a,

$$\frac{L}{H} \sim \frac{1}{12d(\phi)} \left(\frac{H}{a} \right)^2, \quad d(\phi) = \frac{D}{\dot{\gamma} a^2}$$

for achieving steady state in the pressure-driven flow of a viscous suspension, and show that it is considerably longer than $L/H \sim (Re/20)(H/a)$, the channel length required for the boundary layer to reach the centerline in the laminar flow of a homogeneous Newtonian fluid. $d(\phi)$ is a non-dimensional function of the particle volume fraction ϕ and the shear rate $\dot{\gamma}$. It can be estimated from measurements of the self-diffusivity D, for example, $12d(\phi) \sim 1$ for $\phi > 0.3$. The corresponding estimate for the development length L in cylindrical tubes reads as $L/R_0 \sim (R_0/a)^2$. However, experimental evidence published later suggests that the development length ratio may be from one-third to one-half of that predicted by the last equation, Hampton et al. [319]. Similar results were also obtained in the numerical simulations of Phan-Thien and Fang [320]. They argue using this estimate for the flow development length that the disparity in the results of earlier experimental investigations is possibly due to the fact that most of the measurements were taken within the development length before flow reached steady state.

The earliest experimental study of inhomogeneous suspension flow under Stokes flow conditions may have been conducted by Karnis et al. [321] who measured velocity and concentration profiles of neutrally buoyant spheres in a Newtonian fluid in pressure-gradient-driven flow. They reported substantial blunting of the velocity profiles, but no inhomogeneity in the particle concentration; the particles were evenly dispersed in the fluid, as they were at the entrance of the tube. In contrast, the studies of Koh [322] and Koh et al. [323], conducted much later, of pressure-driven flow of a suspension through a rectangular channel using the laser Doppler velocimetry (LDV) technique reveal considerable inhomogeneity in the particle concentration due to migration of particles toward the center of the channel together with blunting of the velocity profile. Other experimental studies have also confirmed the migration of particles from regions of high to low shear rate, Leighton and Acrivos [314], Abbott et al. [324], Hookham [325], and Sinton and Chow [326]. However, the measurements differ in the extent of concentration inhomogeneity and the degree of velocity blunting. Comparison of the Stokesian simulations of Nott and Brady [315] with the experimental findings of Koh et al. [323] is presented in Fig. 6.2.

There are substantial differences between theoretical and experimental findings of concentration and velocity distributions. For low volume concentrations $\phi < 0.2$, simulations overestimate and underestimate, respectively, the concentration distribution in the middle region of the channel and in the region closer to the walls, whereas for $\phi > 0.2$, the reverse is true for the middle region and the agreement is better in the region closer to the walls. The simulations consistently overestimate the blunted velocity profile for all ϕ over most of the channel cross section except closer to the wall. Nott and Brady attribute most of the discrepancies to the experimental measurement flaws like data acquisition within the developing flow region. Among other reasons, they cite, for example, that the method used by Koh et al. [323] was not capable of measuring the bulk density $\phi^b = \int_0^1 \phi \mathrm{d}y$ in Fig. 6.2, and due to the scaling method used to get the bulk density quite possibly the concentration in the channel may have differed from the reported values. In the case of Sinton and Chow [326] who did not detect any migration toward the tube axis, no comparison was possible as the experimental data was for $(H/a) > 200$ outside the range of the simulations of Nott and Brady [315]. However, the data of Sinton and Chow [326] indicates a greater concentration of particles closer to the

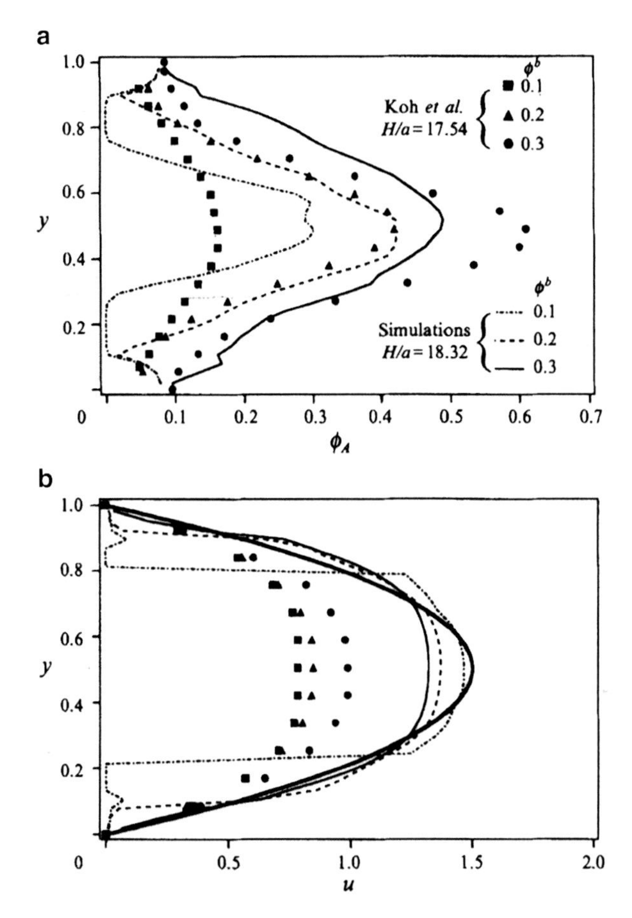

Fig. 6.2 Comparison of the concentration (**a**) and velocity (**b**) profiles of the analytical predictions of Nott and Brady [315] with the experimental results of Koh et al. [323] in the case of the flow of a suspension of monodisperse particles in a channel; H/a is the dimensionless channel width with a the particle diameter; concentrations are given in terms of volume fractions. The *dark line* in (**b**) is the parabolic velocity profile for a Newtonian fluid (reprinted from Nott and Brady [315] with permission)

wall. The discrepancy here is attributed to the relaxation effects of the magnetic resonance imaging (MRI) technique used as well as the data acquisition again performed at a distance from the tube entrance far smaller than that required for reaching steady state.

There are essentially two macroscopic models for suspension flow, the diffusive flux model originally proposed by Leighton and Acrivos [266] to explain their observations of long-time decrease and short-time increase in the effective viscosity of a suspension in a Couette viscometer, and the suspension balance model. Diffusive flux models provide a kinematic description of the shear-induced particle migration in which the particle migration flux is expressed in terms of the gradients of the particle concentration and shear rate. This model was later applied by Phillips et al. [327] to pressure-driven suspension flow in tubes to predict concentration inhomogeneity. In the diffusive flux model, two factors contribute to the net flux of particles: a diffusive flux driven by the gradient of the shear rate [first term in Eq. (6.14)] and diffusion due to a gradient in the concentration for which the diffusivity is proportional to the local shear rate [second term in Eq. (6.14)].

$$\frac{\partial \phi}{\partial t} = \frac{\partial}{\partial y}\left[K_c a^2 \phi \frac{\partial}{\partial y}\left(\dot{\gamma}\,\phi\right) + K_\eta a^2 \dot{\gamma}\,\phi^2 \frac{1}{\eta_s}\frac{\mathrm{d}\eta_s}{\mathrm{d}\phi}\frac{\partial \phi}{\partial y}\right] \tag{6.14}$$

K_c and K_η are proportionality constants, and η_s is the relative viscosity of the suspension. The net flux across the channel is equal to zero at steady state in pressure-gradient-driven flow in a channel reducing the above equation to

$$\frac{\dot{\gamma}\,\phi}{\dot{\gamma}^{w}\phi^{w}} = \left(\frac{\eta_s^{w}}{\eta_s}\right)^{\frac{K_\eta}{K_c}}$$

where w indicates the *wall*. From experimental data on Couette flows, Phillips et al. [327] found that $K_c/K_\eta \sim 0.66$. Using Krieger equation (Eq. 6.11), they arrive at $\phi = \phi_m/(1 + Cy)$, which predicts migration toward the axis but with an unrealistic cusp at the center and no dependence on the ratio H/a.

In the suspension balance model, there are no diffusion considerations; rather mass, momentum, and energy balances are written for the particulate phase and the entire suspension in a manner analogous to molecular or atomic systems. Suspension balance models relate the particle migration flux to the rheology of the suspension. A two-phase approach is used, and from the mass and momentum balance equations of the particle phase and of the suspension as a whole, it is deduced that *the particle migration flux is driven by the divergence of the normal stresses of the particle phase*.

The balance equations are solved simultaneously for the concentration, bulk velocity, and *suspension temperature* a concept introduced by Nott and Brady [315]. The particle positions in a suspension are given by a set of coupled, non-linear ordinary differential equations that gives rise to deterministic chaos. It is this chaotic motion of the particles that is responsible for the shear-induced diffusion in homogeneous shear flow and particle migration in inhomogeneous flows. As a measure of the fluctuating component of the motion of particles in a suspension arising from this chaos, Nott and Brady [315] introduced the *suspension temperature* concept defined as $T = \langle \boldsymbol{u}' \bullet \boldsymbol{u}' \rangle_p$ where $\boldsymbol{u}'$ and $\langle \bullet \rangle$ represent the velocity fluctuation of a particle about its local mean velocity and the angle brackets denote an ensemble average of the inner product over all particles. This scalar definition is appropriate if the magnitude of the velocity fluctuations do not have a directional dependence. If they do, a *tensorial temperature* may be introduced $\mathbf{T} = \langle \boldsymbol{u}'\boldsymbol{u}' \rangle_p$. The fluctuating component of the motion of particles is closely related to the shear-induced diffusivity, thus the appropriateness of the concept of *suspension temperature*. The particles evolve according to Newton's laws of motion just as molecules; the difference is that the forces acting on them are given by rather complex configuration-dependent expressions. In other words, the fluid in this framework only determines the nature of interactions between particles, and once these interactions are known, no explicit reference to the fluid is necessary. Therefore, a macroscopic description of the particles as a continuum should be possible, just as in a molecular system.

The balance equations for the suspension as a whole can be obtained from the conservation of mass and Cauchy's equation of motion by averaging. For an incompressible material,

$$\frac{D\langle \rho \mathbf{u} \rangle}{Dt} = \langle \mathbf{b} \rangle + \nabla \bullet \langle \boldsymbol{\Sigma} \rangle, \qquad \nabla \bullet \langle \mathbf{u} \rangle = 0$$

where $\boldsymbol{\Sigma}$ is the total stress and $\mathbf{b}$ is the body force. The material derivative is now with the suspension average velocity $\langle \rho \mathbf{u} \rangle$. Nott and Brady derive the following low-Reynolds-number constitutive law for the total stress $\boldsymbol{\Sigma}$,

$$\langle \boldsymbol{\Sigma} \rangle = -\langle p \rangle_f \mathbf{1} + 2\eta \langle \mathbf{E} \rangle + \langle \boldsymbol{\Sigma} \rangle_p = -\langle p \rangle \mathbf{1} + 2\eta_f \eta_s(\phi) \langle \mathbf{E} \rangle + \eta \langle \boldsymbol{\chi} \rangle$$
$$\langle p \rangle = \langle p \rangle_f + \Pi, \qquad \eta_s(\phi) = 1 + \eta_p(\phi)$$

where $\langle p \rangle_f$ is the average pressure in the fluid, $\langle \boldsymbol{u} \rangle$ is the average velocity of the suspension (particles and fluid), $\langle \mathbf{E} \rangle$ is its rate of strain, and $\langle \boldsymbol{\chi} \rangle$ represents the averaged normal stresses. Π scales linearly with the shear rate $\Pi = \Pi_0 + \eta \overline{p}(\phi) \dot{\gamma}$ with $\dot{\gamma} = \sqrt{\langle \mathbf{E} \rangle : \langle \mathbf{E} \rangle}$ where (:) represents tensorial inner product. Π_0 is a constant and $\overline{p}(\phi)$ is a monotonically increasing function of the volume fraction. In this suspension balance model, there is no diffusive motion, but there is particle migration from regions of high to low shear rate.

With the introduction of the *suspension temperature* as a fundamental variable determining the particle pressure, a balance law for the fluctuational motion is needed. An equation for the temperature is arrived at by considering the mechanical energy balance for the entire material.

$$\frac{1}{2} \frac{D\left\langle \left(\rho u^2\right)' \right\rangle}{Dt} = \left\langle \mathbf{b}' \bullet \mathbf{u}' \right\rangle + \langle \boldsymbol{\Sigma} \rangle : \langle \mathbf{E} \rangle - \left\langle \dot{\Phi} \right\rangle - \nabla \bullet \langle \mathbf{q} \rangle \tag{6.15}$$

The left-hand side is the rate of accumulation of fluctuational energy, $\langle \mathbf{b}' \bullet \mathbf{u}' \rangle$ is the rate of working by body forces $\mathbf{b}'$, $\langle \boldsymbol{\Sigma} \rangle : \langle \mathbf{E} \rangle$ is the rate of working by the bulk stress (bulk stress is assumed to be symmetric), $\left\langle \dot{\Phi} \right\rangle$ is the average rate of dissipation of mechanical energy into heat, and the last term is the divergence of the heat flux vector $\mathbf{q}$, which is modeled as $\langle \mathbf{q} \rangle = -\langle \boldsymbol{\Sigma}' \bullet \mathbf{u}' \rangle \sim -\eta \kappa(\phi) \nabla T$ based on a Fourier-like argument where $\kappa(\phi)$ is the thermal conductivity proportional to the particle viscosity. The heat flux vector plays an important role in inhomogeneous flows in transporting fluctuational motion to regions of low shear rate and high particle concentration. Note that in low-Reynolds-number flows, the dissipation $\left\langle \dot{\Phi} \right\rangle$ is made up of the dissipation due to the fluid and dissipation due to the particles. The latter is further split into three components, the dissipation due to particles moving with the mean velocity gradient (affine dissipation), dissipation due to all particles moving relative to the mean suspension velocity as will occur when there is phase slip due to a body force acting on the particles (affine dissipation), and dissipation due to the fluctuational motion of the particles about their mean. Through these and similar considerations, Nott and Brady transform the fluctuational energy equation (Eq. 6.15) into

$$\frac{1}{2}c(\phi)\frac{\mathrm{D}\left\langle (\rho u^2)' \right\rangle}{\mathrm{D}t} = \beta(\phi)\left\langle \mathbf{b}' \bullet \mathbf{u}' \right\rangle + \langle \boldsymbol{\Sigma} \rangle_p : \langle \mathbf{E} \rangle - \eta\alpha(\phi)\frac{T}{a^2}\left\langle \dot{\Phi} \right\rangle - \eta\nabla \bullet (\kappa(\phi)\nabla T)$$

where $\alpha(\phi)$, $\frac{\beta(\phi)}{\alpha(\phi)}$ and $c(\phi)$ are non-dimensional phenomenological coefficients representing the temperature from shearing motion, the temperature in sedimentation, and the heat capacity. These macroscopic equations are phenomenologically similar to those used to model the behavior of dry granular flows. However, the origin of the velocity fluctuations is quite different from that in dry granular flows where dissipation occurs within particles due to inelastic collisions. In a fully developed flow in the (x) direction with all the variations occurring in the (y) direction, the above energy balance is reduced to

$$\eta_p(\phi)\left(\frac{\mathrm{d}\langle u \rangle}{\mathrm{d}y}\right)^2 - \alpha(\phi)T + \varepsilon^2\frac{\mathrm{d}}{\mathrm{d}y}\left(\kappa(\phi)\frac{\mathrm{d}T}{\mathrm{d}y}\right) = 0 \tag{6.16}$$

When $\varepsilon^2 = \left(\frac{a}{H}\right)^2 \ll 1$, the above equation simply becomes an algebraic balance,

$$\eta_p(\phi)\left(\frac{\mathrm{d}\langle u \rangle}{\mathrm{d}y}\right)^2 = \alpha(\phi)T.$$

This shows that the concept of "temperature" introduces quadratic non-local effects as it scales as the shear rate squared. Thus, the suspension balance model and the diffusion flux model differ in the sense that the former includes non-local quadratic effects in the shear rate, and the latter depends only on the local shear rate. This implies as well as can be seen from the above equations that the suspension balance model takes into account the flux of fluctuational motion, a non-local effect, while the diffusive flux model does not as it is not set up to account for non-local effects. The flux of the fluctuational motion becomes important in regions where the shear rate is zero and in the boundary layers. Where the shear rate is zero, the conduction of fluctuational motion will result in a finite temperature, *removing the cusp in the density profile that the diffusive model predicts.* Though the small parameter ε multiplies the conduction term in Eq. (6.16), the conduction parameter $\kappa(\phi)$ in Eq. (6.16) diverges as concentration ϕ approaches maximum packing thus keeping the conduction of fluctuational energy finite. It is observed in the Stokesian simulations that in the boundary layers near the walls, there is a rapid variation in T within a layer of thickness ε. The suspension balance model predicts this behavior. It is also straightforward to show that the diffusive flux model is contained within this suspension balance model.

The dimensionless governing momentum balance equations based on the suspension balance model for steady fully developed pressure-driven flow in a channel are

$$\frac{\mathrm{d}}{\mathrm{d}y}\left(\eta_p(\phi)\frac{\mathrm{d}\langle u\rangle_p}{\mathrm{d}y}\right)+1-\varepsilon^{-2}\frac{9}{2}\phi f^{-1}\left(\langle u\rangle_p-\langle u\rangle\right)=0,\quad \frac{\mathrm{d}}{\mathrm{d}y}\left(p(\phi)\sqrt{T}\right)=0. \tag{6.17}$$

with the energy balance the same as Eq. (6.16). This equation immediately predicts without further manipulation that there can be no phase slip between the particles and the fluid as clearly to achieve a balance the following should be satisfied $\langle u\rangle_p=\langle u\rangle+O(\varepsilon^2)$. Then the reduced momentum balance equations (Eq. 6.17) together with the energy balance (Eq. 6.16) can be solved for the unknowns u, T, and ϕ with appropriate boundary conditions.

The concentration profiles predicted by the suspension balance model display a sharp upturn to maximum packing in the $O(\varepsilon)$ thin boundary layers adjacent to the walls, which results from setting the temperature to be zero at the walls. This is not seen in the simulations of Nott and Brady [315]. The sharp upturn in the boundary layer next to the wall can be reduced substantially if the pressure Π is redefined as

$$\Pi=\frac{1}{2}\left[p(\phi)\sqrt{T}+\overline{p}(\phi)\dot{\gamma}\right]$$

Near the centerline where the shear rate $\dot{\gamma}$ vanishes the pressure will be set by $\sqrt{T}$, while near the walls where T is zero, it will be set by $\dot{\gamma}$.

The continuum-based constitutive equation introduced by Phillips et al. [327] couples a Newtonian stress/shear rate relationship with a shear-induced migration model of the suspended particles. The local viscosity of the suspension is dependent on the local volume fraction of solids. In this respect, particle migration due to shear rate, viscosity, and/or concentration gradients are referred to as shear-induced migration. Suspension as a whole is modeled as a single continuum, the motion of which is governed by a non-linear constitutive model for the particle concentration in a flowing suspension.

In the case of neutrally buoyant particles suspended in an incompressible fluid,

$$\rho_o\left(\frac{\partial u_i}{\partial t}+u_j\frac{\partial u_i}{\partial x_j}\right)=\frac{\partial\Sigma_{ij}}{\partial x_j},\quad \frac{\partial u_i}{\partial x_i}=0$$

with $\mathbf{u}$ the mass-averaged suspension velocity and $\mathbf{\Sigma}$ the total stress tensor and ρ_o the suspension density. Phillips et al. [327] assume that the total stress tensor is given by a generalized Newtonian relationship

$$\Sigma_{ij} = -p + 2\eta(\phi)D_{ij}$$

where $\eta(\phi)$ is the effective viscosity of the suspension and $\mathbf{D}$ is the rate of deformation tensor, with $\eta = \eta_f \eta_s$ (see Sect. 6.2.1). The local particle concentration varies with time and must be known in order to evaluate the total stress. The evolution equation for particle volume fraction ϕ representing a balance between stored particles, the convected particle flux, and diffusive particle flux $\mathbf{N}$ is given as

$$\frac{\partial \phi}{\partial t} + u_i \frac{\partial \phi}{\partial x_i} = -\frac{\partial N_i}{\partial x_i}$$

In the absence of Brownian motion and sedimentation, the diffusive particle flux $\mathbf{N}$ is modeled as $\mathbf{N} = \mathbf{N}_\eta + \mathbf{N}_c$ with $\mathbf{N}_\eta$ and $\mathbf{N}_c$ representing the flux contributions due to spatial variations in viscosity and due to hydrodynamic particle interactions, respectively. Phillips et al. [327] model these as

$$\mathbf{N}_\eta = -a^2 \phi^2 \dot{\gamma} K_\eta \nabla(\ln \eta)$$

$$\mathbf{N}_c = -a^2 \phi K_c \nabla\left(\dot{\gamma}\phi\right)$$

where a is the particle radius and K_η and K_c are empirically determined coefficients. Using these equations together with Krieger's formulation for the viscosity, Phillips et al. [314] derive for the distribution of the steady-state particle volume fraction ϕ in the radial geometry of a Couette device,

$$\frac{\phi}{\phi_w} = \left(\frac{r}{R_i}\right)^2 \left(\frac{\phi_m - \phi_w}{\phi_m - \phi}\right)^{1.82\left(1 - \frac{K_\eta}{K_c}\right)}$$

where ϕ_w is the value of ϕ at the inner cylinder and the ratio K_η/K_c is experimentally determined. They conclude that the steady-state particle distribution is independent of particle size. For pipe flow, they derive a similar equation with $\frac{r}{R_i} \rightarrow \frac{R_0}{r}$ where R_0 is the pipe radius. This solution predicts that the particles will migrate toward the pipe centerline where the particle concentration reaches maximum packing ϕ_m with a cusp in particle concentration at the pipe center, which is questionable as was pointed out by Nott and Brady [315] and Koh et al. [323]. Experimental evidence published a few years later indicates that a spike in the particle concentration actually does occur along the pipe axis, but the particle concentration at the pipe center always remains slightly less than ϕ_m for moderate to concentrated suspensions, Hampton et al. [319].

The effect of shear-induced normal stresses in non-colloidal suspensions of rigid particles immersed in a Newtonian fluid of viscosity η_f and density ρ undergoing a set of curvilinear low-Reynolds-number flows which are locally well approximated as simple shear flows such as the wide-gap Couette, parallel-plate torsional, and small angle cone and plate torsional flows had been studied analytically earlier by Morris and Boulay [279] who set out to determine the influence of normal stresses upon the particle fraction and velocity fields for shear flows. They show that shear-induced migrations which have been observed in several curvilinear shear flows of concentrated *monodisperse* suspensions by Phillips et al. [327] and Chow et al. [328] can be explained by the predictions of a rheological model which accounts for the shear-induced particle fraction-dependent, anisotropic normal stresses in Stokesian suspensions $Pe \to \infty$, $Re \to 0$ [see the definitions (Eq. 6.13)]. They did not measure normal stresses; however, they point out that experimental measurements of normal stresses in suspensions were to a large extent incomplete even for the simplest system of a *monodisperse* suspension of non-colloidal spheres. Indeed reliable measurements of normal stresses were done only very recently starting with Zarraga et al. [280]. Their study provides valuable relationships in these curvilinear flows to guide experiments. They examine the role of normal stresses in causing particle migration and macroscopic spatial variation of the particle volume fraction in a mixture of rigid neutrally buoyant (sedimentation is absent) spherical particles for *monodisperse* non-colloidal suspension in Stokes flow.

They state that anisotropic compressive normal stresses are present in sheared suspensions, but their precise form is unknown, Morris and Boulay [279]. A sufficiently simple rheological model that includes normal stresses for analytical steady-state predictions is proposed. The particle contribution to the stress Σ_p is written as

$$\Sigma_p = -\eta_f \dot{\gamma}\,\mathbf{Q}(\phi) + 2\eta_f \eta_p(\phi)\mathbf{E} \tag{6.18}$$

where η_p is the particle contribution to the shear viscosity made dimensionless with η_f and $\mathbf{E}$ is the local bulk suspension rate of strain $\dot{\gamma} = \sqrt{2\mathbf{E}:\mathbf{E}}$. Normal stresses are specified by the dimensionless ϕ-dependent material property tensor $\mathbf{Q}$

$$\mathbf{Q}(\phi) = \eta_n(\phi)(\mathbf{e}_1 \otimes \mathbf{e}_1 + \lambda_2 \mathbf{e}_2 \otimes \mathbf{e}_2 + \lambda_3 \mathbf{e}_3 \otimes \mathbf{e}_3)$$

thus reducing the material response function $\mathbf{Q}(\phi)$ to a single scalar function $\eta_n(\phi)$ and a constant tensor, which describes the anisotropy of the normal stresses in terms of the normal stress anisotropy parameters λ_2 and λ_3 assumed to be positive. The numerical subscripts correspond to the directions in a viscometric shear flow with (1), (2), and (3) indicating the flow, gradient, and vorticity directions, respectively. The function $\eta_n(\phi)$ is called the *normal stress viscosity*. It is made dimensionless with η and is given by the ratio $\left(\eta_n(\phi) = -\Sigma_{11}/\left(\eta\dot{\gamma}\right)\right)$ mirroring the definition of suspension viscosity $\left(\eta_s(\phi) = -\Sigma_{12}/\left(\eta\dot{\gamma}\right)\right)$. Then Eq. (6.18) clearly implies that all normal stresses are negative, and the suspension pressure $\Pi = -(1/3)tr\Sigma_p$ is positive. The constitutive model implies

$$N_1 = -\eta\eta_n\dot{\gamma}(1-\lambda_2), \qquad N_2 = -\eta\eta_n\dot{\gamma}(\lambda_2-\lambda_3)$$
$$\Pi = \eta\eta_n\dot{\gamma}\left(\frac{1+\lambda_2+\lambda_3}{3}\right),$$

with the shear and normal viscosities of the bulk suspension given by Eqs. (6.12) and (6.19), respectively,

$$\eta_n = K_n\left(\frac{\phi}{\phi_m}\right)^2\left(1-\frac{\phi}{\phi_m}\right)^{-2} \tag{6.19}$$

Morris and Boulay [279] determine that $K_s = 0.1$ and $K_n = 0.75$ fit best the wide-gap Couette flow data. They derive a balance equation which represents particle mass–momentum conservation coupling,

$$\frac{\partial\phi}{\partial t} + \langle\mathbf{u}\rangle\bullet\nabla\phi = -\nabla\bullet\mathbf{j}_\perp = -\frac{2a^2}{9\eta_f}\nabla\bullet\left[f(\phi)\nabla\bullet\boldsymbol{\Sigma}_p\right] \tag{6.20}$$

where $\mathbf{j}_\perp$ indicates the flux of particles relative to the mean motion of the suspension, the migration flux, or in other words the cross-stream component of the flux and $\langle\bullet\rangle$ denotes the suspension average of the velocity. The sedimentation hindrance function $f(\phi)$ relates the sedimentation rate of a homogeneous suspension of spheres at volume fraction ϕ to the isolated Stokes settling velocity usually expressed in a form $f(\phi) = (1-\phi)^\alpha$ first suggested by Richardson and Zaki [329]. Morris and Boulay [279] use $\alpha = 4$ in their computations. The full particle flux and the component perpendicular to the main stream are given by

$$\mathbf{j} = \phi\langle\mathbf{u}_p\rangle = \phi\langle\mathbf{u}\rangle + \frac{2a^2}{9\eta_f}f(\phi)\nabla\bullet\boldsymbol{\Sigma}_p$$
$$\mathbf{j}_\perp = \phi\langle\mathbf{u}_p\rangle - \phi\langle\mathbf{u}\rangle$$

where $\langle\mathbf{u}_p\rangle$ is the local average velocity of the particle phase. It should be noted that the fluid viscosity η_f sets the scale for the particle stress Σ_p for a given shear rate $\dot{\gamma}$. When $\nabla\bullet\boldsymbol{\Sigma}_p = 0$, migration ceases and all particles are advected with the bulk motion. Through simple examples, Morris and Boulay [279] demonstrate that Eq. (6.20) indicates that *normal stresses* are responsible for the cross-stream flux of particles. The similarity of this physical mechanism with the *normal stress-driven migration of polymer chains* in solution, studied by MacDonald and Muller [330], among others, is remarkable. However, the modeling fails to predict a realistic concentration distribution at the centerline in channel flow, where the bulk average shear rate vanishes. The remedy is either to introduce the concept of *suspension temperature* in the analysis, Nott and Brady [315], or a non-local normal stress approach based upon formation of a particle network as proposed by Mills and Snabre [331].

The bulk suspension balance, for neutrally buoyant Stokes flow suspensions, satisfies $\nabla \bullet \Sigma = 0$ at all times. During the period of migration beginning from onset of flow of a uniformly concentrated suspension, the phase stress divergences are both non-zero but are locally equal and opposite $\nabla \bullet \Sigma_p = -\nabla \bullet \Sigma_f$ with stress gradients developing in both phases of the flowing mixture, driving the components in opposite directions and resulting in bulk segregation. When steady state is reached, $\nabla \bullet \Sigma_p = 0$ is satisfied. Using this equation, Morris and Boulay [279] compute in steady-state parallel-plate torsional flow $\eta_n(\phi) = A_1 r^{(1-2\lambda_3)/\lambda_3}$, in steady-state wide-gap Couette flow $\eta_n(\phi) = A_2 r^{(1+2\lambda_2)/\lambda_2} \eta_s(\phi)$, and in steady cone and plate flow $\eta_n(\phi) = A_3 r^{(1+\lambda_2-2\lambda_3)/\lambda_3}$. In deriving these relationships, the possibility of slip at the wall of the particle phase is neglected and a no-slip condition at the wall is assumed. All constants A_i, $i = 1, 2, 3$ are determined by requiring ϕ to sum to ϕ_{bulk}. The first relationship predicts that the relationship between ϕ and r is independent of the magnitude of the normal stresses. The experiments of Chow et al. [328] indicate that migration in the parallel-plate geometry is weak or non-existent implying a constant η_n, which can be achieved in the present modeling by setting $\lambda_3 = 0.5$.

The complex behavior of suspensions of neutrally buoyant spheres in a Newtonian liquid was modeled numerically by Subia et al. [332] via a Galerkin, finite element formulation incorporating the continuum diffusive flux model of Phillips et al. [327]. This shear-induced migration model is shown to be surprisingly robust at capturing the essential features in two-dimensional flows, such as concentric Couette flow and cylindrical tube flow, and three-dimensional axisymmetric suspension flows, such as flow in a journal bearing (counter-rotating eccentric cylinders) and piston-driven flow in a pipe. The evolution of the solid concentration profiles of initially well-mixed suspensions subjected to slow flow is determined non-invasively using the nuclear magnetic resonance imaging in the experiments to benchmark the finite element code. Good qualitative and quantitative agreement of the numerical predictions and the experimental measurements are presented.

6.2.4 Challenges in Shear-Driven Migration of Suspensions

The field of suspension flow prediction in general flows other than shear flows is wide open and has not been tackled as yet. In most practical cases, the suspension will be subjected to flows other than simple shear, and a constitutive equation appropriate for these flow situations is lacking. There are substantial remaining difficulties in simple shear flows as well. For example, the ratios $\lambda_2 = \frac{\Sigma^p_{22}}{\Sigma^p_{11}}$ and $\lambda_3 = \frac{\Sigma^p_{33}}{\Sigma^p_{11}}$, where the superscript p means "particle," are taken as constant. However, these ratios should be ϕ dependent, and an understanding of this dependence is required. Taking them as constants may succeed in describing the phenomena qualitatively, but quantitative prediction may/will require knowledge of the ϕ dependence. Two other issues which seem to be directly relevant to most applications are the prediction of the flow of *bidisperse* suspensions and more difficult yet *polydisperse* suspensions and a *viscoelastic* suspending medium rather than a Newtonian one. All the reported simulations and predictions are for *monodisperse* suspensions except for Krishnan et al. [333] who discuss experimental evidence for radial particle segregation in the parallel-plate geometry. The tracer particles are seen to experience a constant drift velocity independent of their radial position. They point out that in their experiments, tracer particles migrate inward or outward in the parallel-plate flow if they are smaller or larger, respectively, than the other particles of an otherwise *monodisperse* suspension contrary to expectations based on the experiments of Abbott et al. [324] who had observed that larger particles migrated radially outward to regions of lower shear stress in a wide-gap Couette device. There is no work in the archival literature on the total particle stress and shear-induced diffusion of polydisperse systems although shear viscosity of multimodal suspensions has been studied. It is worthwhile to remark that prominent descriptions of microstructure, such as the pair distribution function, are not easily applied to polydisperse systems. In addition, the maximum packing fraction possible ϕ_m for a given suspension composition and packing arrangement, which is an isotropic measure of microstructure, varies with polydispersity. A more complete description of suspension microstructure for polydisperse systems to be used in constitutive models is not available at this time.

The physical mechanism embedded in the suspension balance models requires that inhomogeneous particle normal stresses drive particle migration with the migration flux governed by the divergence of the normal stress of the particle phase, Nott and Brady [315], Mills and Snabre [331], and Morris and Boulay [279]. But the exact nature of the particle stress that is responsible for migration is still subject to debate. L'huillier [334] and Nott et al. [335] argue that only the component of the particle normal stress that comes from *non-hydrodynamic* interactions between particles may be involved in migration phenomena. When that stress-induced migration exists, it is always in concurrence with the shear-induced migration. There exist very few experimental or numerical data on the values of the particle normal stresses and on their variation with particle concentration.

L'huillier [334] reaffirms that migration is the result of two physically distinct processes. A shear-induced process first suggested by Leighton and Acrivos [278] and a stress-induced process first suggested by Nott and Brady [315]. Much remains to be understood about the migration process as most if not all numerical simulations based on Stokesian dynamics, have dealt exclusively with the particle stress equation. Sierou and Brady [285] had previously found that the *hydrodynamic* stress displays different normal stresses, non-Newtonian effects related to the anisotropic microstructure of the suspension that is to particle arrangements with preferred directions. Based on these findings, L'huillier [334] proposed a closure (constitutive relationship) for the particle stress by introducing a new variable to represent the anisotropy at the microstructural level and borrowing its evolution equation from non-linear viscoelastic fluid mechanics with modifications, in his case the CFD equation of Criminale et al. [114]. The role of the shear-induced microstructure in setting up the anisotropic properties of the suspension is gaining more emphasis and attention as demonstrated by the recent research of Blanc et al. [336] who conducted experiments for particle volume fractions ranging from 0.05 to 0.56 with transparent suspensions made of polymethyl-methacrylate particles of 172 μm in diameter dispersed in a fluorescent index-matched Newtonian liquid sheared in a wide-gap Couette rheometer. Their experimental findings are in good agreement with Stokesian dynamics simulations where the interaction force between particles has been tuned to reproduce the particle roughness effects.

The complex nature of concentrated suspensions $\phi > 20$ % and sometimes contradictory behavior reported in the literature makes it challenging to construct a theoretical model. Most of our current understanding of non-colloidal suspensions comes from experimental and numerical observations. Numerical simulations in particular greatly contribute to our understanding of concentrated suspensions in ways not always available in experiments. Stokesian dynamics method, Brady and Bossis [316, 317], Sierou and Brady [284, 285], and Brady [337], has been widely used for simulating the unique behavior of dense suspension flow at low-particle Reynolds number, Dratler and Schowalter [338] and Drazer et al. [339]. There are other simulation techniques such as lattice Boltzmann method, Chen and Doolen [340] and Hill et al. [341], and the Lagrange multiplier fictitious domain method, Glowinski et al. [342], Singh et al. [343]; however, the Stokesian dynamics method in which the linear equations describing Stokes flow are simultaneously solved at discrete time steps for all the particles in the simulation seems to be the most widely used for a wide range of volume fractions. Stokesian dynamics simulations of concentrated suspensions have been mostly applied to flows in unbounded domains, and as a consequence, our understanding of such flows is now much more advanced than the behavior of suspensions in wall-bounded domains where the presence of a solid boundary makes the characteristics of the suspension dramatically different. There have been a few numerical studies of the wall-bounded suspensions using the Stokesian dynamics simulations, where a wall is replaced with a chain of fixed spheres notably by Nott and Brady [315] and Singh and Nott [289]. But these simulations have been limited to a monolayer simulation of a small number of particles of the order $O(10)$ due to the high computational cost. A few three-dimensional simulations of the Couette flow have been performed using the lattice Boltzmann method, Nguyen and Ladd [344] and Kromkamp et al. [345]. Of particular note are the results of the last authors who showed that the wall structuring in two-dimensional simulation differs from the three-dimensional results. They also observed dependency of some results on the computational resolution. The shortcomings of the Stokesian dynamics method in terms of simulating a small number of particles have been overcome to a large extent by the force-coupling method developed by Maxey and Patel [346] and Lomholt and Maxey [347]. The method computes the far-field hydrodynamic interaction between particles by solving the Stokes equations and can simulate a large number of particles, up to 10^4 particles in suspension, Dance and Maxey [348]. However, the method does not represent the exact solution if the distance between two particles is closer than $r/a < 2.4$ in which r and a denote the particle center-to-center distance and the particle radius, respectively, Lomholt and Maxey [334], which proves to be adequate for low volume fractions. Consequently most force-coupling simulations have been performed in the semidilute regime $\phi < 20\ \%$. In concentrated suspensions, strong short-range lubrication forces are generated between particles in close proximity as fluid in the intervening gap is displaced by the relative motion of the particles. These forces, together with near-surface contact forces, play an important role in the suspension rheology and self-diffusion of particles. However, these forces also lead to ill-conditioned problems for determining the particle stresses and particle motion in large systems of particles at higher volume fractions. Recently, Yeo and Maxey [349] developed a more robust lubrication correction method for the simulation of near-field effects in concentrated suspensions to remedy these problems and gave examples of numerical simulations with more than 4,000 non-Brownian, spherical particles in a homogeneous shear flow. Subsequently, they extended the method to wall-bounded flows to investigate the effects of the wall on the suspension rheology by including modifications to take into account the particle–wall lubrication interaction, Yeo and Maxey [350]. The fully three-dimensional simulations of concentrated suspensions $0.2 \leq \phi \leq 0.4$ were performed with particles of $O(10^3)$ in a Couette flow at zero Reynolds number for $10 < L_y/a < 30$ where L_y and a represent the channel height and the particle radius, respectively. Their results are among the first to elucidate the structure of the suspension flow in bounded domains. They find that the strong particle–wall lubrication inhibits particles near the wall from being resuspended into the core region. Thus, one or more particle layers are formed near the wall depending on the volume fraction and the suspension can be divided into three regions depending on the microstructures: the wall region where a structured particle layering is dominant, the buffer region which shows the characteristics of both the particle layer and the shear structure, and the core region in which the suspension field is quasi-homogeneous. The width of the inhomogeneous region (wall and buffer) is a function of ϕ and not sensitive to L_y/a. They tentatively extend their findings to all wall-bounded flows such as Poiseuille flow citing evidence of particle layering that exists in the literature, Hampton et al. [319].

The measurements of the first and second normal stress differences of an extremely shear-thickening polymer dispersion were reported by Laun [351]. The sample was a polymer dispersion of spherical particles of 280 nm average diameter in glycol. In the regime of strong shear thickening, a negative first normal stress difference N_1 is found. The data obtained from cone and plate rheometry shows that $N_1 \approx -|\tau| \approx -2N_2$, where τ is the shear stress, prompting Laun to conclude

$$\Sigma_{11} = -\frac{1}{2}|\tau|, \Sigma_{22} = \frac{1}{2}|\tau|, \Sigma_{33} = 0$$

without accounting for the suspension pressure Π. Morris and Boulay [279] point out that it is difficult to reconcile the tensile stress Σ_{22} with experimental evidence and argue that suspension pressure should be included in the interpretation of the data,

$$\Sigma_{11} = -\left[\frac{\Pi}{3} + \frac{1}{2}|\tau|\right], \Sigma_{22} = -\left[\frac{\Pi}{3} - \frac{1}{2}|\tau|\right], \Sigma_{33} = -\frac{\Pi}{3}$$

with all stresses compressive if $\Pi > \frac{3}{2}|\tau|$. That all components of the normal stress should be compressive (negative) in concentrated suspensions ($0.45 < \phi < 0.52$) of near-hard spheres has been established by the findings of Yurkovetsky [307] with $|\Sigma_{33}| < |\Sigma_{22}| < |\Sigma_{11}|$.

6.2.5 Particle Motion in Viscoelastic Suspending Media at Very Low Reynolds Numbers

The modeling of the normal stresses in the mixture of a viscoelastic fluid and neutrally buoyant particles is complicated because normal stress contributions arise from both particles and fluid. There are very few investigations reported in the archival literature on this issue. Tehrani [352] demonstrated experimentally the effect of a viscoelastic suspending medium. He reports experiments on particle migration in viscoelastic fluids used in hydraulic fracturing. His results can be summarized as follows: particle migration is controlled by the elastic properties of the suspending fluid and the shear rate gradient. In fluids with low but measurable normal stresses and dominant shear-thinning properties, particles migrate to regions of lower shear rate. Migration is fast initially, but slows down rapidly over a short distance. For these fluids, the bulk migration velocity correlates with the product of the Weissenberg number and the mean shear rate gradient. In contrast, highly elastic fluids with relaxation times well above one second and shear-thickening properties at low shear rates, flow with a central plug region or slip at the wall, producing little or no migration.

Before Tehrani, the focus of most investigations in the literature was on the effects of shear-thinning generalized Newtonian fluids or the effects of elasticity without shear thinning. As early as the early 1960s, the migration of spheres in a pipe flow was studied by Karnis and Mason [353] neglecting inertial effects. They observed that neutrally buoyant spheres migrate toward the center region of lower shear rate in a viscoelastic fluid with constant viscosity but toward the pipe wall in a purely shear-thinning fluid. That the particles migrate toward the region of highest shear rate in very slow flows of pseudoplastic fluids was confirmed by Gauthier et al. [354]. The elastic and viscous effects on migration of neutrally buoyant solid spherical particles at low volume solid fraction in plane Poiseuille flow was investigated experimentally under creeping flow conditions by Jefri and Zahed [355] who observed that the particles are in uniform distribution in a Newtonian fluid (a pure corn syrup) but migrate toward the centerline in a Boger fluid (0.2 wt% polyacrylamide in corn syrup) resulting in the formation of a narrow core. The class of viscoelastic fluids exhibiting normal stresses and elasticity but very little shear thinning is labeled Boger fluids in honor of their discoverer.

Stokesian dynamics simulations as described by Brady and Bossis [317] were modified and extended by Binous and Phillips [356] to the case of non-shear-thinning viscoelastic suspending media (Boger fluids). Stokesian dynamics simulations work by accounting for the far-field interactions through the use of multipole moment expansions, and near-field lubrication interactions between particles are accounted for by using a pairwise additivity approximation. The suspending visco-elastic fluid is modeled as a suspension of finitely extensible non-linear elastic (FENE) dumbbells (see section 2.3.2.2 in Siginer [73]) in a Newtonian solvent. Dumbbell suspensions in their simulations show negligible shear-thinning, thus the characterization of the suspending fluid as a Boger fluid. However, instead of averaging to obtain the well-known FENE model, a continuum constitutive equation, particle-to-particle and particle-to-bead interactions are calculated directly by using a modified version of the Stokesian dynamics method. In their modified Stokesian dynamics simulation, particle motion is computed directly from a kinetic theory description of the viscoelastic fluid under low-Reynolds-number conditions, without the need for a pre-averaged constitutive equation and without the need for discretizing space. Thousands of bead-and-spring dumbbells can be included in the simulation. The interactions between bead–sphere as well as between spheres are calculated by using relations from low-Reynolds-number hydrodynamics. In principle, the method can be applied to N spheres. However, Binous and Phillips [356] present only one and two particle simulations, experimental data, and theoretical information for which is available in the literature. To determine if the velocity fields with and without dumbbell–dumbbell interactions are close enough or significantly different, they conduct simulations in each case, calculate the velocity profiles with and without dumbbell interaction contributions, and compare contour plots to find that the dumbbells collectively do *not* cause a significant modification of the overall velocity field if the dumbbell–dumbbell interactions are taken into account. Thus, not accounting for dumbbell–dumbbell interactions *does not result* in a velocity field in which dumbbells are deforming under the influence of an inappropriate velocity field one in which each dumbbell is responding to a velocity field contributed by the spheres as if no other dumbbells were present. In all their simulations with non-spherical particles and with two spherical particles, the velocity contours with and without dumbbell–dumbbell interaction contributions look almost the same. This finding is consistent with the well-known fact that velocity fields of Boger fluids often exhibit very small deviations from the Newtonian field, Lunsmann et al. [357]. However, they also found significant differences in the velocity fields with one sphere simulations, thus confirming the importance of accounting for dumbbell–dumbbell interaction in computing the velocity field in this particular case.

Their findings in *unbounded* flow domains include the significant drag increase on a single sphere by the presence of dumbbells. In high Deborah numbers, the velocity of sedimenting sphere becomes time periodic. Earlier, Walters and Tanner [358] had suggested that elasticity causes a slight drag reduction at low Deborah (*De*) numbers, but causes a marked increase in drag as the *De* increases. The experiments of Solomon and Muller [359] confirmed the conjecture of Walters and Tanner [358]. A substantial increase in the drag with increasing *De* numbers was shown with a threefold increase for Deborah numbers greater than unity, the largest increase occurring for Boger fluids with the greater extensibility. Numerical computations of Chilcott and Rallison [360] and Harlen [361] in that order had already suggested the possibility of drag enhancement in Boger fluids. Binous and Phillips [356] numerical computations confirm previous findings.

They also confirm computationally previous findings of experimental studies on the sedimentation of non-spherical particles in viscoelastic fluids in the absence of inertia; non-spherical particles rotate as they fall such that their long axis is ultimately pointed in the direction of gravity, Tiefenbruck and Leal [362], Chiba et al. [363], and Liu and Joseph [364]. The comprehensive experiments of the last authors demonstrate that the tilt angles of long cylinders and flat plates falling in viscoelastic liquids are determined by the competition between viscous effects, viscoelastic effects, and inertia. They differentiate between shape tilting and inertial tilting. Viscoelasticity dominates when inertia is small and the particles settle with their broadside parallel or nearly parallel to the direction of fall. However, cylinders with round ends and cone ends tilt much less when inertia is small (slow flow). Normal stresses acting at the corners of rectangular plates and squared-off cylinders with flat ends cause shape tilting from the vertical. Particles settle with their broadside perpendicular to the direction of fall when inertia is large. The tilt angle varies continuously from 90° when viscoelasticity dominates to 0° when inertia dominates. Particles will turn broadside-on when the inertia forces are larger than viscous and viscoelastic forces. This orientation occurs when the Reynolds number Re is less than 0.1 in Newtonian liquids and very dilute solutions. Long particles will eventually turn their broadside perpendicular to the stream in a Newtonian liquid for any $Re > 0$, but in a viscoelastic liquid, this turning cannot occur unless $Re > 1$. Another condition for inertial tilting is that the elastic length λU should be longer than the viscous length ν/U where U is the terminal velocity, ν is the kinematic viscosity, and $\lambda = \nu/c^2$ is a relaxation time where c is the shear wave speed. The condition $M = U/c > 1$, where M represents the viscoelastic Mach number (see Chap. 4 on Transcriticality for the definition of the Mach number), is interpreted as a hyperbolic transition of solutions of the vorticity equation analogous to transonic flow. Strong departure of the tilt angle from $\theta = 90°$ begins at about $M = 1$ and ends with $\theta = 0°$ when $1 < M < 4$. Much earlier, Kim [365] had shown analytically that axisymmetric particles tend to settle in the direction of their axis of rotation in the case of the suspending fluid abiding by the second-order CE.

The first study of a pair of spheres sedimenting in a viscoelastic fluid goes back to the 1970s. That two sedimenting spheres will come together, rotate to a vertical orientation, and fall as a doublet in a second-order fluid was shown analytically by Brunn [366]. About the same time, Riddle et al. [367] found experimentally that two spheres sedimenting in a tube move together or apart, depending on whether their initial separation is less or greater than a critical value. More than a decade and a half later, Joseph et al. [368] showed experimentally that, during sedimentation at low Reynolds number, groups of spheres in polymer solutions tend to cluster together into rows oriented in the direction of motion. Two sedimenting spheres are in most cases attracted to each other, and the line through the centers of the resulting two particle configuration when they come together side by side initially rotates to line up with the direction of gravity. Experimental observations in the literature indicate that the rotational velocity of the line through the centers of the spheres from the side-by-side configuration is in the same direction as it would be in a Newtonian fluid. Gheissary and van den Brule [369] reported surprising observations of anomalous behavior on time effects during sedimentation in shear-thinning fluids (Carbopol) which on a time scale of hours shows no relaxation of the wake structure; in fact, the opposite occurs, and time effects seem to become stronger. Their experiments with two, or more, settling particles indicate that the interaction between particles is complicated by the time effects in the wake of the leading particle. Phillips [370] describes a new method, which requires the calculation of only the low-Reynolds-number Newtonian velocity profile, to compute the motion of N spherical particles suspended in a quiescent second-order fluid. The Newtonian profile is used to evaluate particle velocities accurate to leading order in the Deborah number. Simulation results involving two, three, four, and six particles are reported as illustrative examples as well as the formation of clusters in the sedimenting system, which are in qualitative agreement with experimental observations.

The computations of Binous and Phillips [356] are in keeping with the experimental observations on record. It is important to note that their method clearly reveals the effects of elasticity of the suspending medium; however, the method does not include the effects of inertia and shear thinning and is restricted to low Reynolds numbers and constant shear viscosities. In a follow-up paper, they extend their approach to problems with *bounded* geometries, Binous and Phillips [371], by using the image system of Blake [372] to examine particle–wall interactions in suspensions of FENE dumbbells. They find that the lateral velocity of a single sphere sedimenting near a wall is proportional to its fall velocity as the sphere moves toward the wall and exhibits anomalous rotation. Their data also suggests the existence of a critical sphere–wall separation beyond which attraction becomes negligible. The particle–wall attraction is strong enough to overcome a slightly tilted wall, but the particle moves away from the wall when it is tilted by an angle larger than a critical value. Sphere weight and size have a small and large effect, respectively, on the distance traveled by the sphere before touching the wall. They show that a neutrally buoyant sphere in plane Poiseuille flow migrates to the channel center and to the walls when the sphere size is sufficiently small (wide

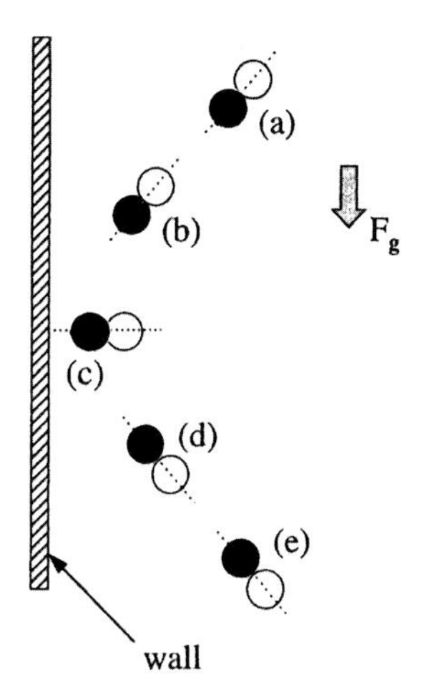

Fig. 6.3 Turning of a non-spherical particle as it approaches a wall in a viscous fluid (reprinted with permission from Binous and Phillips [371])

channels) and large (narrow channels) relative to the channel width, respectively, that is, the migration direction depends on the ratio of sphere size to channel width. The migration rate is larger at higher Deborah numbers and for larger spheres. Finally, a non-spherical particle falling between two walls in a Newtonian fluid at low Reynolds numbers has an oscillatory trajectory. Viscoelasticity dampens these oscillations, causing the particle ultimately to position itself in the center of the channel with its long axis parallel to gravity. A non-spherical particle made up of two spheres in a Newtonian fluid released from a location far from either wall will drift toward one wall in a direction determined by the orientation of its long axis relative to gravity. As it approaches one of the walls, sphere–wall lubrication interactions slow the sphere nearest the wall, and the pull of gravity on the second (upper) sphere causes it to fall, reorienting the particle such that it begins to drift toward the opposite wall (Fig. 6.3).

Particle moves back and forth between the two walls repeating the motion indefinitely in a Newtonian fluid, resulting in oscillatory trajectories. The behavior of a non-spherical particle is qualitatively different in a viscoelastic fluid represented by dumbbells suspended in a Newtonian solvent. After a few oscillations, the particle positions itself in the center of the channel and falls with its long axis in the direction of gravity. The particle centering is more rapid at higher Deborah numbers or higher dumbbell concentrations. All of these predictions are consistent with experimental observations published in the literature, Liu and Joseph [364] and Joseph and Liu [373], indicating that the method outlined may provide a promising approach for simulating a wide class of flows involving particles suspended in viscoelastic fluids. Similar results were also obtained by Huang et al. [374] who did two-dimensional calculations of the motion of a circular particle in plane Poiseuille flow.

To study the effect of volume fraction ϕ of solids on flow behavior, Huang and Joseph [375] work with volume fractions of 0.21 and 0.42 in Newtonian, generalized Newtonian, and shear-thinning Oldroyd-B type suspending medium.

They determine that practically particle-free zones which exist in more dilute cases at the basically shear-free channel center in Newtonian and generalized Newtonian fluids are suppressed in concentrated suspensions. The migration away from the center in more dilute cases is caused by the curvature of the velocity profile and is diminished with increasing volume fractions, Asmalov [376].

6.2.6 *Secondary Field in Poiseuille Flow of Shear-Driven Migration of Suspensions*

The effect of the normal stress difference that generated secondary currents on the concentration distribution in the cross plane in concentrated suspension flow in conduits of non-circular cross section was addressed for the first time by Ramachandran and Leighton [377] who showed that when modeling concentrated suspension flows in non-circular tubes, it is critical to consider the rheology of the concentrated suspension to arrive at the correct steady-state particle distribution profile. In the absence of both secondary currents and wall-slip effects, the steady concentration distribution of a suspension flowing through a straight conduit of arbitrary cross section is independent of the particle size. But Ramachandran and Leighton [377] have shown that the steady-state particle concentration distribution is no longer independent of particle size in concentrated solutions (more than 10 % phase volume) as it is assumed for dilute solutions but it depends on the aspect ratio B/a with B and a representing the characteristic length scale of the cross section and the suspended particle radius, respectively, as well as on the shear-induced migration Péclet number Pe, the ratio of the shear-induced diffusive time scale to the convective time scale, which scales as B^2/a^2 and which characterizes the shear-induced migration velocity. Péclet number Pe is also interpreted as the ratio of the transverse convective velocity to the shear-induced migration velocity. Particle migration is dominated by secondary currents if this ratio is larger than 10^2.

The constitutive equation for diffusive particle flux N of Phillips et al. [327] for shear flows in two dimensions assumes that particle migration is isotropic.

$$N = -K_c a^2 \left(\phi^2 \nabla \dot{\gamma} + \dot{\gamma} \phi \nabla \phi \right) - K_\eta \frac{\phi^2}{\mu} \frac{\partial \mu}{\partial \phi} \dot{\gamma} a^2 \nabla \phi \tag{6.21}$$

Here, $\dot{\gamma}$ is the local shear rate, a is the particle radius, and μ is the viscosity of the suspension as a function of the volume fraction ϕ with K_c and K_η representing experimentally determined volume fraction-dependent phenomenological constants. The structure of Eq. (6.21) indicates that particle migration across the streamlines is governed by viscosity, shear, and concentration gradients. Viscosity gradients push the particles from regions of higher viscosity to regions of lower viscosity. This equation has severe shortcomings in spite of some of its apparent successes. For instance, it cannot explain the migration of particles from high curvature areas of conduits of non-circular cross section to lower curvature areas. It actually predicts the opposite. It cannot predict the apparent lack of migration in the parallel-plate flow in which the shear rate increases radially outward and the outward migration of particles in cone and plate flow in which the shear stress is constant.

To remedy these deficiencies, the anisotropy of the particle stress tensor should be taken into account. Ramachandran and Leighton [377] show that this approach results in a curvature term directly proportional to the second normal stress difference coefficient, the local particle pressure, and inversely proportional to the local radius of stream surface curvature in the particle flux balance equation. This term, which is responsible for the sharper particle migration toward the tube center in anisotropic suspensions, also represents the forcing function for the secondary currents and establishes a force that drives a flow along the local radial vector normal to the axial stream surfaces. The weak secondary flows are linked to the concentration distribution equation by a convective term that scales as the shear-induced migration Péclet number. For circular geometries, this driving force is invariant in the azimuthal θ direction, and therefore, there is zero net flow in the cross section. However, when there is a lack of symmetry in the cross section, as would occur with the introduction of an eccentricity to the circle, this term produces a net force that drives circulation currents. Since the driving force is directly proportional to the local curvature, the anisotropic particle stress terms drive a stronger flow from the side regions of the ellipse than from the top and bottom of the ellipse. This results in a non-zero secondary current that flows in from the side regions along the major axis of the ellipse and flows back along the minor axis and the top and bottom walls.

The origin of the secondary currents in the flow of suspensions through conduits at steady state is attributed to the asymmetry caused by an azimuthal gradient in the curvature of the flow stream surfaces such as occurs in any non-axisymmetric flow. This source of asymmetry is weak and leads to small magnitudes of the circulation currents. Therefore, the effect of these currents is realized only at high Péclet numbers. The concentration contours and the secondary velocity profiles for the flow of a 40 % suspension in an elliptical channel of aspect ratio $W = 2$ (length scale B is the semi-minor axis of the ellipse) for the isotropic model and the anisotropic model with two values of the Péclet number $Pe = 0$ and $Pe = 1{,}600$ show that shear-induced migration results in the diffusion of particles from the top and bottom of the tube toward the midplane (major axis) of the ellipse. The second normal stress induces a convected flow from the high-curvature regions near the sides toward the center. At high $Pe = 1{,}600$, the combination of these effects results in a significant depletion of particles from the side regions. Particles are convected from the sidewalls toward the center of the ellipse more strongly relative to the isotropic and the $Pe = 0$ concentration profile. The concentration predicted by the isotropic model near the sidewall is around 37 %, while that predicted by the anisotropic model at $Pe = 1{,}600$ is just 27 %. The depletion near the sidewall with respect to the 40 % flow average concentration is thus nearly four times that predicted by the isotropic model. The concentration and secondary current profiles in a rectangular channel of aspect ratio $W = 2$ (length scale B used to define Pe is the half-depth of

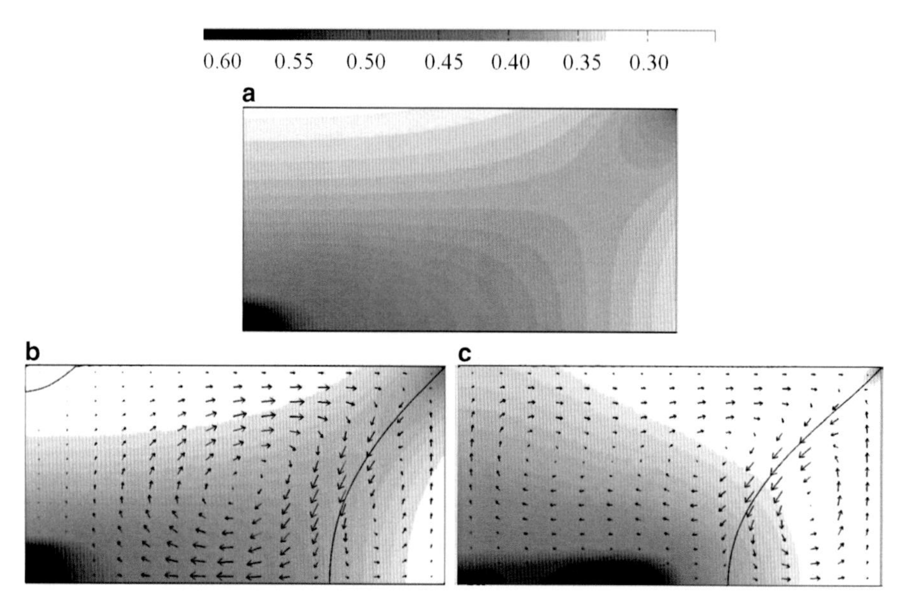

Fig. 6.4 Concentration and secondary current profiles for a rectangular channel with an aspect ratio of 2 at 40 % bulk concentration for (**a**) $\chi = 0$, (**b**) $\chi = 400$, and (**c**) $\chi = 1{,}000$. The profiles are shown over one-quarter (first quadrant) of the cross section (reprinted with permission from Ramachandran and Leighton [377])

the channel) are shown in Fig. 6.4. Similar to the case of the ellipse, the concentration profiles are shifted away from the sidewalls toward the center of the channel for the anisotropic model relative to the isotropic model.

The similarity between the secondary motions induced in viscoelastic polymer flow and suspension flow in tubes of non-circular cross sections is striking. In both cases, the magnitude of the secondary circulation is inherently weak, typically two to three orders of magnitude lower than the axial velocity, but in both cases, they cause drastic changes in the flow structure. In the pressure-gradient-driven suspension flow through conduits of arbitrary cross section second normal stress difference-induced secondary currents drastically affect the steady-state particle distribution; although very weak, they prove to be the dominant mechanism governing the particle concentration distribution, and they lead to counterintuitive results. For instance, the isotropic shear-induced migration argument alone yields a concentration profile, which is the exact opposite of the concentration profile that would be obtained if the effect of the secondary flows in the cross section was accounted for. The isotropic shear-induced migration approach alone predicts that particles should migrate into corners in a square or rectangular geometry, for instance, but if secondary currents are taken into account, particles are actually flushed out of the corners instead of accumulating in the corners. The secondary flow effects are most likely to be observed when the average concentration ranges from 30 to 50 %, when the magnitude of the currents is high relative to the shear-induced migration velocity for reasonably large Péclet numbers Pe.

The existence of secondary flows in the pressure-driven flow of a concentrated suspension of non-colloidal particles through a conduit of square cross section under creeping flow conditions has been recently confirmed experimentally, Zrehen and Ramachandran [378]. Their research lends support to the idea that secondary currents, rather than shear-induced migration, may actually be the dominant mechanism that determines particle distribution in non-colloidal suspension flows through non-axisymmetric geometries. Their work also establishes that co-extrusion of two concentrated suspensions through non-axisymmetric geometries with a stable suspension–suspension interface is not possible, except in special situations.

6.2.7 Secondary Field in Single-Phase Turbulent Flow of Suspensions

In fully developed single-phase turbulent flow in straight pipes, it has been known that mean motions can occur in the plane of the pipe cross section, when the cross section is non-circular or when the wall roughness is non-uniform around the circumference of a circular pipe. This phenomenon is known as secondary flow of the second kind and is associated with the anisotropy in the Reynolds stress tensor in the pipe cross section. In a similar way in particle-laden fully developed turbulent flow of Newtonian fluids even in smooth circular tubes, secondary flow of the second kind can be promoted by a non-uniform non-axisymmetric particle forcing, which makes the Reynolds stress tensor anisotropic, Belt et al. [379]. Small enough droplets in suspension would behave in the same way as solid particles, and much that can be said about a suspension of solid particles also applies to a suspension of small droplets.

Extending Eqs. (6.5) and (6.6) and adding the forcing $\mathbf{\Phi}$ due to the particles in the cross section,

$$\rho \frac{\mathrm{D}\mathbf{u}}{\mathrm{D}t} = -\nabla \tilde{p} + \eta \Delta \mathbf{u} + \nabla \bullet \tilde{\boldsymbol{\tau}} + \mathbf{\Phi}, \qquad \nabla \bullet \mathbf{u} = 0$$

$$\rho \frac{\mathrm{D}\zeta}{\mathrm{D}t} = \eta \Delta \zeta + \nabla \wedge \nabla \bullet \tilde{\boldsymbol{\tau}} + \nabla \wedge \mathbf{\Phi}$$

where $\mathbf{\Phi}$ is the projection onto the cross section of the average forcing per unit volume $\langle F \rangle$ with $\mathbf{\Phi} = \langle F \rangle - (\langle F \rangle \bullet \mathbf{e}_z)\, \mathbf{e}_z$ where $\mathbf{e}_z$ is the unit vector in the axial direction. Belt et al. [379] further split $\mathbf{\Phi}$ into two components, a component independent of the secondary flow, $\mathbf{\Phi}_o$, and another dependent on the secondary flow, $\mathbf{\Phi}_v = \mathbf{\Phi} - \mathbf{\Phi}_o$, $\mathbf{\Phi}_v \equiv 0$ if $\mathbf{u} = 0$. In turbulent flows, if the forcing is proportional to the fluid velocity (in the case of a fixed distribution of particles with linear drag), the forcing will be large but $\mathbf{\Phi}_o = 0$. If the drag is non-linear as would be the case with moving particles, $\mathbf{\Phi}_o$ may not be necessarily equal to zero and may induce a secondary flow. In the former case, the necessary and sufficient condition for the existence of secondary flow remains $\nabla \wedge \nabla \bullet \widetilde{\boldsymbol{\tau}} \neq 0$. But in the latter case, the necessary and sufficient condition reads as $\nabla \wedge \nabla \bullet \widetilde{\boldsymbol{\tau}} + \nabla \wedge \boldsymbol{\Phi} \neq 0$. Belt et al. [379] perform laser Doppler anemometry experiments to show that a non-uniform internal forcing on the flow as represented by particles distributed non-uniformly in the cross section can promote a secondary flow. In turbulent pipe flow, particle forcing due to non-uniform particle distribution promotes changes in the turbulence resulting in an anisotropic Reynolds stress tensor in the cross section, which triggers the secondary flow. They show the occurrence of a particle-driven secondary flow consisting of four symmetrical cells (for the particular particle distribution used in the experiments). Significantly, the magnitude of the secondary flow velocity can be as high as 9 % of the bulk velocity u_B. Secondary flow pattern is determined by the dominant terms in the divergence of the Reynolds stress tensor $\widetilde{\boldsymbol{\tau}}$ in the cross section, the radial and azimuthal gradients of the radial and circumferential Reynolds stresses. The divergence of the projection $\widetilde{\boldsymbol{\tau}}$ onto the cross section of the Reynolds stress tensor reads as

$$(\nabla \bullet \widetilde{\boldsymbol{\tau}})_r = \frac{\partial \tau_{rr}}{\partial r} + \frac{1}{r}\frac{\partial \tau_{r\theta}}{\partial \theta} + \frac{\tau_{rr} - \tau_{\theta\theta}}{r}$$

$$(\nabla \bullet \widetilde{\boldsymbol{\tau}})_\theta = \frac{1}{r}\frac{\partial \tau_{\theta\theta}}{\partial \theta} + \frac{\partial \tau_{r\theta}}{\partial r} + 2\frac{\tau_{r\theta}}{r}$$

$$\tau_{\theta\theta} = -\rho\overline{u'_\theta u'_\theta}, \qquad \tau_{rr} = -\rho\overline{u'_r u'_r}, \qquad \tau_{r\theta} = -\rho\overline{u'_r u'_\theta}$$

The gradients of $\tau_{r\theta}$ is of much smaller magnitude than the gradients of τ_{rr} and $\tau_{\theta\theta}$ in the cross section leading to the conclusion that particle forcing does not have a large impact on $\tau_{r\theta}$ and by extension on its gradient. This is in contrast to single-phase turbulent flow in non-circular cross sections where the gradient of the Reynolds shear stress $\tau_{r\theta}$ in the cross section plays a major role in the determination of the secondary flow, Demuren and Rodi [251]. All this should be contrasted as well with single-phase turbulent flow without particle forcing in round tubes in which case $\tau_{r\theta} \equiv 0$.

The overall momentum balance is affected by the turbulence dynamics modified by the wake of the particles. Belt et al. [379] developed a scaling law that gives the magnitude $|\mathbf{u}|$ of the velocity $\mathbf{u}$ of the secondary flow within the range of the bulk Reynolds number Re_B in their experiments, 5,300 > Re_B > 10,600, with u_B, N_P, D_P, and D standing for the magnitude of the longitudinal bulk velocity, the particle density, the particle diameter, and the tube diameter, respectively.

$$\left(\frac{|\mathbf{u}|}{u_B}\right)^2 \approx \frac{3\pi}{4} N_P D_P D^2 \frac{1}{Re_B}\left[1 + 0.15\left(\frac{D_P}{D}\right)^{0.687}(Re_B)^{0.687}\right]$$

Particle density in the experiments was $N_P \sim 1.5 \times 10^6/\text{m}^3$.

Gas–solid flows in wall-bounded confined systems occur frequently in industrial transport processes like pneumatic conveying, fluidized beds, vertical risers, and flow mixing devices, among others. Pneumatic conveying is widely used in the chemical, food processing, and cement industries and in thermal power plants. It may be categorized into different regimes depending on dust mass loading and transport velocity such as dilute phase conveying, stream-type conveying, and dense phase conveying. The particle mass loading is defined to be the ratio of the mass of the particles to the mass of the gas in the computational domain. Small suspended particles can change the structure of the turbulence if they are present at sufficient concentrations. In turbulent gas flow, three regimes of particle mass loading, m_s, can be identified: $m_s \ll 1$ (dilute phase), $m_s \sim O(1)$ (stream type), and $m_s \gg 1$ (dense phase). In the former case, the effect of the particles on the gas phase is not expected to be significant. In the other cases, the effect may be significant; however, the particle volume fraction can still be very small compared to unity prompting to ignore the volume occupied by the particles in simulating the gas flow. Criteria governing the tendency of the particles to suppress gas-phase turbulence provided the average particle size is sufficiently small were suggested by Hetsroni [380] and Gore and Crowe [381]. The former investigated the interaction between solid particles and the turbulence of the carrier fluid and suggested based on theoretical considerations supported by available experimental data that particles with low Reynolds number cause suppression of the turbulence, while particles with higher Reynolds number cause enhancement of turbulence due to wake shedding. The significance of the particle diameter to fluid length scale ratio in determining whether or not the addition of a dispersed phase would cause an increase or decrease in the carrier phase turbulent intensity was discussed by Gore and Crowe [381].

Dilute phase conveying is in extensive use in coal-based power plants to transport fine pulverized coal from the mill to the burners. Typical coal dust loadings are in between 0.2 and 0.6 (kg dust/kg air) with the conveying velocity not exceeding 20 m/s. It is important to establish a homogeneous coal dust distribution within the pipe cross section in front of distributors or manifolds in order to guarantee that the different burners are supplied with the same amount of coal. Further, the coal dust should be homogeneously distributed in the supply pipe entering the burner for an efficient combustion with low pollution. The segregation of the coal dust from the conveying air usually occurs as a result of gravitational settling in horizontal pipes or inertial effects in pipe bends. A reduction of the transport velocity could yield considerable energy savings. But that would require a detailed understanding of particle behavior in turbulent pipe flows. The particle motion in wall-bounded gas–solid flows is influenced by gravitational settling in horizontal pipes, inertial behavior in pipe bends and branches (Huber and Sommerfeld [382]), turbulent dispersion (Lee and Durst [383]), transverse lift forces (i.e., Magnus effect) induced by particle rotation which is mainly caused by particle–wall collisions (Sommerfeld [384]), lift forces due to shear flow (Saffman [385] and Mei [386]), the collision of the particles with the rough walls of the pipe (Matsumoto and Saito [387] and Sommerfeld [388]), wall collision process for non-spherical particles (Tsuji et al. [389]), and particle-to-particle collisions (Oesterle and Petitjean [390]).

Dilute phase conveying where the particles are fully dispersed in the flow and no deposition occurs has been investigated by Huber and Sommerfeld [391] who performed three-dimensional numerical simulations of particle-laden flows based on the Euler/Lagrange approach for the fluid and dispersed phases, respectively, including two-way coupling, turbulence, turbulent particle dispersion, particle interactions with rough walls, and collisions between particles. There had been previous attempts to simulate such flows; however, theirs seems to be the first published simulation that takes into account all major physical interactions that influence the flow without introducing strongly simplifying assumptions. It should be noted that unless the simulations are three dimensional, a true picture of the flow will not emerge. That is because even the horizontal pipe flow is already a three-dimensional problem due to the gravitational settling of the dust and the resulting modification of the gas flow.

The numerical calculations were performed by the Euler/Lagrange approach for the fluid and dispersed phases, respectively. The k-ε closure model was used with the time-averaged Navier–Stokes equations, Launder and Spalding [392]. Lagrangian approach is used to deal with the solid phase, which implies that particles are followed in time along their trajectories through the flow field. A collection of particles with the same properties were grouped in so-called 'parcels' in order to allow the representation of the concentration of the dispersed phase by a reasonable number of computational particles. The added mass effect and the Basset history force have been neglected in the present calculations, since a gas–solid two-phase flow with a large density ratio ρ_P/ρ was considered. The Reynolds stress tensor is clearly anisotropic because of the non-uniform distribution of the particles (based on the study of Belt et al. [379]). However, no secondary flows would be generated because the isotropic k-ε closure model has been used. But secondary flows emerge nevertheless due to the higher collision frequency of particles in the bottom section of the pipe as compared with the upper section as a result of gravitational settling, which gradually increases the particle density toward the bottom. In addition, the effect of wall roughness translates into a transfer of particle momentum from the axial direction to a direction normal to the wall; that is, the average rebound velocity normal to the wall becomes larger than the normal component of the impact velocity. Also the curvature of the wall refocuses the trajectories of the bouncing particles toward the pipe centerline. The combination of all these effects results in a locally higher momentum transfer to the gas phase in the bottom section generating a secondary flow. Huber and Sommerfeld [391] compute the maximum velocity of this secondary flow to be of the order of 2 % of the average conveying mainstream velocity. They also observe that a secondary flow is not detected with smooth pipes.

Li et al. [393] performed numerical simulations of particle-laden flows down vertical channels with much larger particle mass loadings, up to 2, accounting for both elastic and inelastic collisions between particles. They conclude that it makes little difference whether the collisions are elastic or inelastic and that particle–particle collisions significantly reduce the tendency of particles to accumulate near the walls of the channel. Particle–particle collisions also seem to play a significant role in suppressing the gas-phase turbulence. For the largest mass loading considered, the gas-phase turbulence is strongly suppressed by the particles. For small mass loadings, the particles produce an increase in the mean gas velocity, whereas for the largest mass loading, the reverse is true. For small mass loadings, there is a decrease in the fluctuation intensities of the normal and spanwise components of velocity. The particles have a large effect on the velocity–pressure-gradient terms in the turbulence energy balance that inhibits the transfer of energy from the streamwise component of velocity to the other two components of velocity resulting in greater anisotropy in the turbulence.

6.2.8 Secondary Field in Multiphase Turbulent Flow of Suspensions

The existence of horizontal, annular two-phase flow has been well documented, but the physical mechanism underlying its occurrence was still controversial until the mid-1990s when Flores et al. [394] put to rest the ongoing controversy at that time concerning the contribution of secondary flows to maintaining two-phase gas–liquid annular flow in tubes and even its existence. They unequivocally show that it exists and its contribution can no longer be discounted. Until 1995, mechanisms have been proposed but none can be decisively supported with data. Flores et al. [394] present direct measurements of secondary flow together with semi-empirical correlations to both prove the existence of gas-phase secondary flow and to predict annular to stratified transition limits for isothermal and heated horizontal annular two-phase flows in pipes. They show that circumferential secondary flow in the vapor gas core exists, and they hypothesize that to be the predominant cause of the annular flow in the transition region between annular and stratified flows where the gas core flow rates are below or near the onset of entrainment. Prior to Flores et al. [394], the effects of the secondary flow on the liquid film were quite controversial. Jayanti et al. [395] argued the tangential shear of the secondary flow to be insufficient to sustain a liquid film at the top of the circular tube, and Jayanti and Hewitt [396] even questioned the existence of secondary flows at about the same time as Flores et al. [394]. However, the measurements of Flores et al. [394] together with those of Dykhno et al. [397] and Williams et al. [398] firmly established the existence of secondary flows.

Horizontal heated tubes are extensively used in once-through boilers and fluidized bed combustors. The use of horizontal tubes does have inherent advantages such as greater stability due to the absence of flow rate fluctuations triggered by pressure drop inherent in down flow. However, they do suffer from asymmetric liquid film distribution, which may become a major operational problem as it precedes "dry-out" of the liquid film in the upper half of the tube. "Dry-out" is a condition that happens when part of the pipe wall is not covered with the liquid film. It may lead to corrosion of the wall and deterioration of the heat exchange from the

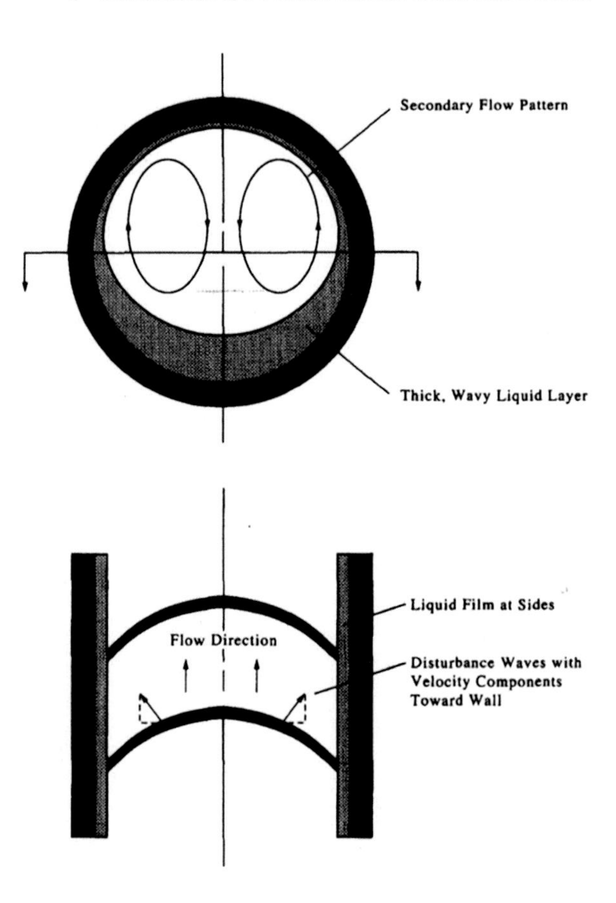

Fig. 6.5 Secondary flow and wave-spreading mechanisms in two-phase gas-liquid flow in a horizontal tube (reprinted with permission from Flores et al. [394])

wall to the liquid. In the stratified "dry-out" regime, the stratified–annular boundary is not complete. This boundary has been theoretically predicted by Taitel and Dukler [399].

The liquid layer is thinnest on the top in annular flow; due to gravitational pull, the liquid film at the bottom is significantly thicker than that on the sides or the top of the tube, and it is quite wavy due to imposed shear of the high-speed vapor core. It is well known that the speed of gravity waves in still water increases with depth. In addition in a moving liquid, the fluid velocity also contributes to the overall wave speed. Both the depth and the fluid velocity are maximum at the bottom of the tube in horizontal annular flow. Due to the circumferentially varying depth and fluid velocity, wave fronts in the thicker bottom layer do not display a uniform speed. As a result, wave crests oriented normal to the tube axis at the very bottom get tilted and have a gradually growing component toward the wall as the wall is approached (Fig. 6.5).

With the force of the flowing gas on the wave, liquid can impact the pipe wall and be transferred up the wall in the same way that water moves up onto the beach as gravity waves approach the shore. This is known as wave spreading and is generally accepted to aid in producing annular flow. Indeed in flow visualization experiments, Flores observed that is the case in his unpublished SM thesis. However, in no cases did any of the waves transport liquid beyond the midplane of the tube, that is, wave spreading did not progress past the midplane of the tube.

Thus the liquid film at the bottom is thicker and also rougher than at the top of the pipe as seen by the gas flow. A turbulent pipe flow experiencing a variation of roughness along the pipe wall will show a secondary flow consisting of two counter-rotating cells in the cross section of the tube, Hinze [250]. Also it is known that a top to bottom circumferential pressure gradient exists in asymmetrically roughened pipes, and it is linked to the generation of secondary flows. However, it seems doubtful that small circumferential pressure gradients could transport liquid against gravity. At high gas velocities, Jayanti et al. [358] measured the time-dependent behavior of annular films in a horizontal pipe as a function of flow conditions and circumferential angle. They conclude that, at high gas rates, there is circumferential spreading of liquid from the bottom. It is believed that the secondary flow in the gas core occurs as the result of the tilting and stretching of vortex rings and the anisotropy of the Reynolds stresses. The former is the cause of Prandtl's secondary flow of the first kind, which also occurs, for example, when a laminar flow goes around a bend. The latter generate Prandtl's secondary flow of the second kind, a strictly turbulence phenomenon. Hinze [250] concludes on the basis of a simplified turbulent kinetic energy equation (Eq. 6.2) that there must be an interchange of fluid from wall zones to the core, in particular along the bisectrix toward the corners in the case of non-circular ducts and along the normal toward a rough wall, as in fully developed flow in the region next to the wall turbulent kinetic energy production exceeds viscous dissipation. The corners of a duct of rectangular cross section and the area next to a rough wall are regions of high turbulence production. In annular flow, the pipe bottom where the liquid carpet is thick and wavy is such a "rough" region. Secondary flow is toward this carpet and back up the pipe walls as shown in Fig. 6.5.

Flores et al. [394] conclude that "The secondary flows are a result of both the circumferential roughness variation of the liquid film, as witnessed by the gas flow, and the asymmetric shape caused by the same film thickness variation. Any roughness variation, step or continuous, will cause this secondary flow. At low gas velocity the major factor which transports liquid into the upper half of the tube causing the transition from the stratified to annular flow is the circumferential secondary flow in the gas core. At higher gas velocities, the deposition of entrained liquid is a more significant factor in transporting liquid to the top of the tube."

Van't Westende et al. [400] investigate numerically, using Eulerian–Lagrangian large-eddy simulations (LES), how the secondary flow, induced by a variable wall roughness, influences the dispersed-phase distribution and the droplet deposition rate in the horizontal annular dispersed pipe flow, with air and water the gas and liquid phases, respectively, which occurs often in pipe transportation of gas and oil and in heat exchangers. In this type of flow, the liquid-phase flows partly as a thin film along the tube wall and partly as entrained droplets in the turbulent gas core.

Due to gravitational pull, the liquid film at the bottom is thicker and rougher than at the top of the pipe, and the concentration of drops is also higher in the bottom region of the pipe than in the top region. A turbulent pipe flow experiencing a variation of roughness along the wall as seen by the gas in the core on the surface of the annular film will show a secondary flow, which consists of two counter-rotating cells in the cross section of the tube. The transversal field alters the distribution of the droplets inside the pipe and their deposition at the wall. Van't Westende et al. [400] show that the presence of secondary flow increases the droplet concentration in the core of the pipe and the droplet deposition rate at the top of the pipe.

An important parameter for this flow regime is the film flow rate along the wall. When and if dry-out happens, that is, part of the pipe wall becomes exposed and is not covered with a liquid film, corrosion of the wall can take place together with a marked deterioration in the heat exchange rate from the wall to the liquid. As the gravitational pull on the liquid film results in a constant drainage of liquid from the top of the pipe toward the bottom to maintain a liquid film along the entire wall of the pipe, that is, to prevent dry-out, other physical mechanisms are to be present to transport the drained liquid back to the top of the pipe. A number of such mechanisms have been proposed over the years. For air–water pipe systems larger than 5 mm in diameter, the main mechanisms are (1) secondary gas flow: a mean flow in the cross section of the pipe, usually manifested as counter-rotating cells, which drags the liquid film upward through the action of the tangential shear force it applies to the surface of the liquid film. Secondary flow can be induced by a varying wall roughness on the surface of the liquid film, by the non-uniformity of droplet concentration or by the fact that the gas is flowing through a non-circular cross section. (2) Entrainment/deposition: droplets, mostly atomized from the thick film at the bottom part of the pipe, can deposit downstream in the top region, where they contribute to the film. (3) Wave spreading: large amplitude waves, being deformed by the non-uniform depth of the liquid film, tend to bend sideways and spread over the circumference. Lin et al. [401] concluded that secondary flow and entrainment/deposition are the dominant mechanisms for the film distribution.

The droplets are also driven by the turbulence of the core gas flow. The turbulent drag force tends to push them toward zones of low turbulence intensity, that is, toward the walls. This phenomenon known as turbophoresis, Young and Leeming [402], deposits the droplets uniformly along the wall, provided the turbulence intensity shows cylindrical symmetry. The combination of secondary flows and turbophoresis affects the distribution of the droplets and their deposition at the wall and thus has an indirect effect on the liquid film. Gravitational settling, turbulence interactions, and secondary flow are thus the processes that drive the dispersed phase (Fig. 6.6).

Van't Westende et al. [400] solve the turbulent gas core with a bulk velocity of 20 m/s and a pipe diameter of 5 cm. The thin liquid film is modeled as a cylindrical wall with a varying wall roughness; the bottom wall being rough, $k_s/D = 0.03$, and the top wall of the pipe being smooth, $k_s/D = 0$. The dispersed phase is simulated using monodispersed solid spheres, with a diameter of either 50 or 100 μm, driven by drag and gravity. For comparison, simulations with uniform roughness and with

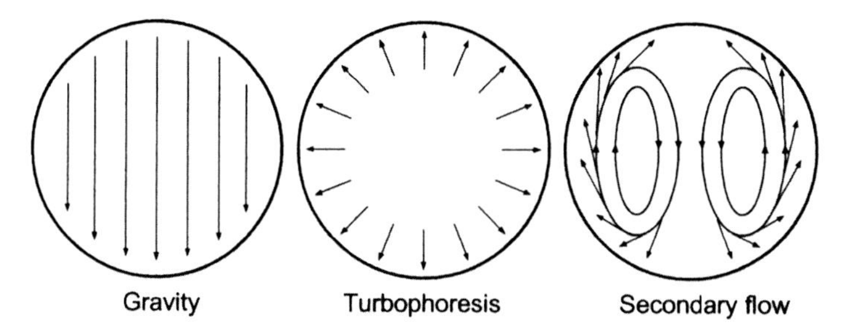

Fig. 6.6 Particle deposition mechanisms in a horizontal annular dispersed pipe flow (reprinted with permission from Westende et al. [400])

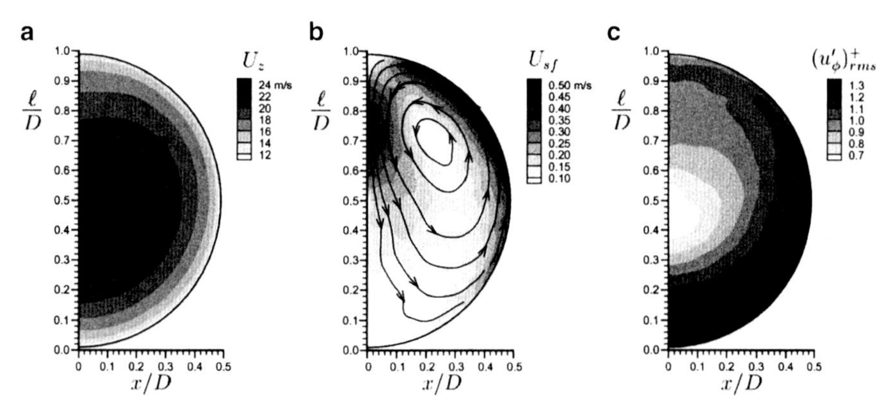

Fig. 6.7 Contour plots of simulation: (**a**) mean axial velocity U_z, (**b**) secondary-flow U_{sf} streamlines, and (**c**) tangential velocity fluctuation $(u'_\varphi)^+_{rms}$ (reprinted with permission from Westende et al. [400])

a smooth wall are also performed. In horizontal particle-laden pipe flows, the particles are pushed toward the wall by the turbophoresis and toward the bottom by the gravity. This leads to a quick depletion of particles (droplets) in the core of the pipe. The secondary flow induced by the variable film thickness can have a large effect on the particle distribution and their deposition at the wall. The major global effect of the secondary flow is to bring the particles from the wall region to the core of the pipe and, in particular, to the top part of the pipe. This increase in the particle concentration in the core and top of the pipe leads to an overall increase in the rate of the deposition of the particles, which also becomes more uniform over the wall circumference, with the rate of deposition at the top becoming of the same order of magnitude as the rate of deposition at the bottom.

Some simulation results are presented in Fig. 6.7. The maximum value of the mean axial velocity is not in the center of the pipe, but is shifted toward the rough bottom. This is the result of the secondary flow (downward through the center of the pipe), bringing axial momentum from the center of the pipe to the bottom region. It thus seems as if the secondary flow "pushes" the position of maximum axial velocity downwards. The secondary flow is determined by the pattern of the Reynolds stresses in the cross section of the pipe. At the bottom of the pipe, the wall shear and the Reynolds stresses are larger (Fig. 6.7c). The gradient of the tangential Reynolds stresses along the wall, shown in Fig. 6.7c, pushes the flow in the opposite direction from the high toward the low values of tangential velocity fluctuation $(u'_\phi)^+_{rms}$ creating the secondary flow pattern shown in Fig. 6.7b.

Stratified flow at fairly low flow rates, which occurs in the gas–liquid flow through a nearly horizontal rectangular duct with the heavier fluid flowing along the bottom, manifests itself in two different regimes: the stratified smooth and stratified wavy regimes in fully turbulent flow. In the latter regime, a strong mean secondary flow with a two-roll cellular structure each in the liquid phase and in the gas phase with the liquid flowing up near the walls and down in the middle of the duct and the gas flowing in reverse order up in the middle and down next to the walls was observed by Suzanne [403]. The wavy regime with secondary flows was also observed later in circular tubes. That the secondary velocities in the liquid phase may result from an interaction between wave pseudomomentum P_z per unit mass and mean axial vorticity ζ_z was shown by Nordsveen and Bertelsen [404]. They use the Lagrangian mean theory of Andrews and McIntyre [405], the mean flow momentum equations of whom include an additional "pseudomomentum" or "wave momentum" term to modify Eq. (6.1) to show that the axial vorticity is given by the following:

$$\zeta_z = \left(\frac{\partial P_z}{\partial x} \frac{\partial W}{\partial y} - \frac{\partial P_z}{\partial y} \frac{\partial W}{\partial x} \right)$$

in which the first term describes the interaction of the wave pseudomomentum P_z across the duct with the vertical variation of the axial velocity and the second term represents the interaction of the vertical variation of the pseudomomentum P_z with the variation of the axial velocity across the duct (Figs. 6.8 and 6.9). Both fields, wave pseudomomentum and axial velocity, decrease with depth. The lateral variation of the wave pseudomomentum is due to a varying wave field amplitude, which was measured by Suzanne [403].

Numerous streaks parallel to the wind direction may be observed when wind blows over an open reservoir of liquid and generates waves. That these streaks are convergence lines between counter-rotating vortices below the surface was

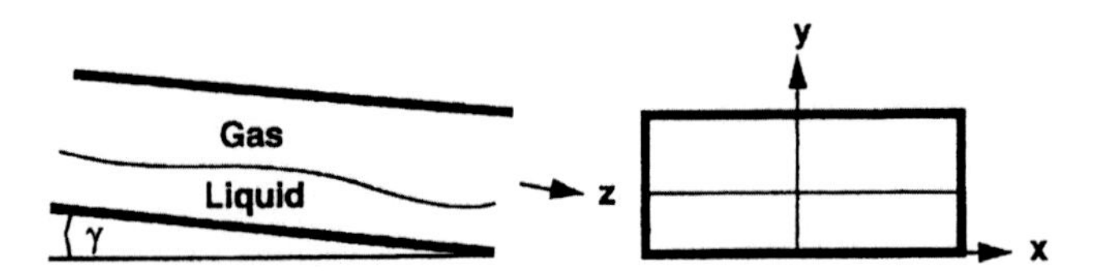

Fig. 6.8 Longitudinal and transversal cross sections of the duct

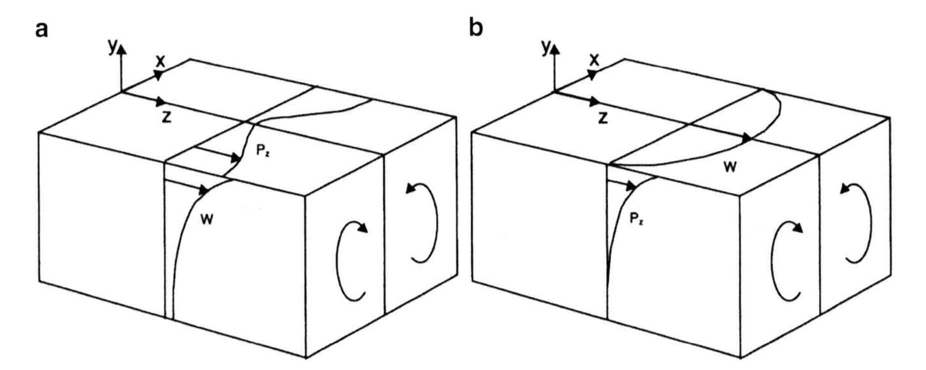

Fig. 6.9 Sketch of the entities involved in the wave pseudomomentum—shear flow vortex forces: (**a**) wave pseudomomentum P_z across the duct with the vertical variation of the axial velocity, (**b**) vertical variation of the pseudomomentum P_z with the variation of the axial velocity across the duct (reprinted with permission from Nordsveen and Bertelsen [404])

discovered by Langmuir [406] in 1938. The vortices carry his name. In the same way, Langmuir type of vortices appears in two-phase gas/liquid duct flow. However, there is an essential difference between flow say over a lake and two-phase flow in ducts in that in the former the wind-induced shear stress is relatively weak and the wave field may be treated as irrotational. But in ducts, there is a strong shear flow and the wave field is rotational. Nordsveen and Bertelsen [407] developed a rotational wave solution with a wave field composed of two crossing linear wave trains interacting with a strong mean shear flow. The magnitude of the secondary currents induced by this wave field was in fairly good agreement with the measurements of Suzanne [403]. However, there were discrepancies between predicted and measured values of the secondary circulation, which differed significantly in particular close to the walls.

Epilogue

Significant progress has been made in the last half of a century in predicting the longitudinal and transversal steady as well as quasi-steady velocity fields of non-linear viscoelastic fluids in straight tubes of arbitrary shape. A unified field theory for viscoelastic in particular non-linear viscoelastic fluids along the lines of the Navier–Stokes equations for Newtonian fluids is still lacking. Several constitutive structures popularly used are subject to Hadamard and dissipative instabilities. The derivation of others violates fundamental continuum mechanics and thermodynamic principles. Naturally they have a bearing on the ability of the constitutive formulations in predicting flows in general geometries and for high Weissenberg *We* numbers. However, in spite of these general difficulties with viscoelastic fluid flow computations, the longitudinal field in arbitrary cross-sectional tube flows for both generalized Newtonian and viscoelastic fluids can be predicted quite well with existing constitutive formulations, but that is not the case for the transversal field.

Hadamard type of instabilities, which lead to blow-up in numerical computations, and dissipative instabilities may preclude the use of many popular constitutive equations at high Deborah *De* numbers common in materials processing industry. However that being said the existing plethora of equations in use does provide good predictions sometimes even quantitative predictions at low *De* numbers in experimental laboratory settings and for dilute fluids if the use of a particular constitutive equation (CE) is restricted to a class of motions. For instance, the well-known upper convected Maxwell model is quite good in predicting elongational flows at low elongation rates, but it is totally useless in predicting secondary flows. A plethora of CEs that may be stable in the Hadamard and dissipative sense and are compliant with the principles of thermodynamics predict a zero second normal stress difference. The search for a universal CE for non-linear viscoelastic fluids has led so far down many blind alleys, and it is by no means certain that success in this respect will manifest itself in the near future.

The progress in predicting the flow of non-colloidal suspensions in straight tubes has also been significant although major developments occurred only over the last three decades or so; however, available theories at this time fall quite short of

D.A. Siginer, *Developments in the Flow of Complex Fluids in Tubes*,
DOI 10.1007/978-3-319-02426-4

drawing a complete and reliable picture. The first particle-based numerical simulations have been performed only two decades ago, and the merits of the competing shear-induced migration-based and continuum-based theories to predict concentration distribution in monodispersed non-colloidal suspensions are still under discussion. Polydispersity is a problem that has not been investigated as yet. The fact that reliable normal stress measurements in monodispersed suspensions have been performed only in the last few years by taking and adapting the methods and ideas developed decades ago by researchers in viscoelastic flows such as using free surface deformations caused by normal stresses of which the classical rod climbing (rod dipping in suspensions) and inclined trough flows are prominent examples really sets the stage for further substantial advances. But it also shows that the tools available to us at this time have not matured to the extent of the predictive powers of the theories for the flow of viscoelastic fluids are. Another example to give support to this statement is the first calculation of the secondary flows of non-colloidal suspensions in tubes of non-circular cross section driven by normal stresses performed only in 2008.

It is self-evident that secondary flows of particle-laden fluids in turbulent regime be it single phase or multiphase cannot be understood and tackled directly without an in-depth understanding of the turbulent secondary flows of linear fluids in tubes. Progress in this area has been substantial, and a comprehensive summary concerning the computation of secondary flows of linear fluids in turbulent tube motions is given in this work. As a final remark we note that this monograph is far more comprehensive and goes well beyond the recent reviews, Siginer [408, 409].

Appendix A
Non-linear Viscoelastic Constitutive Equations and Secondary Flows

Viscoelastic fluids which are constitutively described by the following mathematical structures will not develop secondary flows and will display rectilinear particle pathlines when flowing in tubes of cross section other than circular. In keeping with the continuum mechanics tradition in the literature, the extra-stress tensor is denoted by $\mathbf{S}$.

- *Upper convected Maxwell (UCM) model*: $N_2\left(\overset{\bullet}{\gamma}\right) = 0$.

$$S_{kl} + \lambda \overset{\nabla}{S}_{kl} = 2\eta_o D_{kl},$$

$$\overset{\nabla}{\mathbf{S}} = \frac{\mathrm{D}\mathbf{S}}{\mathrm{D}t} - \nabla\mathbf{u}^{\mathrm{T}}\mathbf{S} - \mathbf{S}\nabla\mathbf{u} = \overset{\bullet}{\mathbf{S}} - \mathbf{DS} - \mathbf{SD},$$

  where $\mathbf{S}$, $\mathbf{D}$, $\mathbf{u}$, λ, and η_o stand for the extra-stress tensor, the rate of deformation tensor, the velocity field, the relaxation time, and the zero shear rate viscosity, respectively. The superscript T and the filled circle on top of the tensor indicate the transpose and the material derivative, respectively. It follows that

$$\eta\left(\overset{\bullet}{\gamma}\right) = \eta_o, \qquad N_1\left(\overset{\bullet}{\gamma}\right) = 2\lambda\eta_o\left(\overset{\bullet}{\gamma}\right), \qquad N_2\left(\overset{\bullet}{\gamma}\right) = 0.$$

  where η is the viscosity, and N_1 and N_2 represent the first and the second normal stress difference.

- *Oldroyd-B model*: $N_2\left(\overset{\bullet}{\gamma}\right) = 0$.

$$S_{kl} + \lambda \overset{\nabla}{S}_{kl} = 2\mu\left(D_{kl} + \lambda_r \overset{\nabla}{D}_{kl}\right),$$

D.A. Siginer, *Developments in the Flow of Complex Fluids in Tubes*,
DOI 10.1007/978-3-319-02426-4

This model is capable of predicting elastic behavior but not a shear-dependent viscosity. $\mathbf{S} = \mathbf{S}_p + \mathbf{S}_s$, the polymer stress $\mathbf{S}_p$ satisfies the UCM equation with $\eta = \eta_p$, and the solvent $\mathbf{S}_s$ stress satisfies $\mathbf{S}_s = 2\eta_s\mathbf{D}$. It follows that

$$\lambda_r = \frac{\eta_s \lambda}{\eta_s + \lambda_p}, \quad \eta = \eta_p + \eta_s$$

$$\eta = \mu, \qquad N_1\left(\overset{\bullet}{\gamma}\right) = 2\mu(\lambda - \lambda_r)\overset{\bullet}{\gamma}, \qquad N_2\left(\overset{\bullet}{\gamma}\right) = 0.$$

- *White–Metzner equation*: $N_2\left(\overset{\bullet}{\gamma}\right) = 0$.

$$S_{kl} + \lambda(\mathrm{II}_\mathbf{D})\overset{\nabla}{S}_{kl} = 2\eta(\mathrm{II}_\mathbf{D})D_{kl}, \quad \mathrm{II}_\mathbf{D} = 2\mathbf{D} : \mathbf{D}$$

The dependence of the relaxation time λ on the second invariant II_D of the stretching tensor $\mathbf{D}$ introduces shear thinning,

$$\lambda(\mathrm{II}_\mathbf{D}) = \frac{\lambda_0}{1 + a\lambda_0(2\mathbf{D} : \mathbf{D})^{1/2}}$$

$$\eta = G\lambda(\mathrm{II}_\mathbf{D}), \qquad N_1\left(\overset{\bullet}{\gamma}\right) = 2G[\lambda(\mathrm{II}_\mathbf{D})]^2\overset{\bullet}{\gamma}, \qquad N_2\left(\overset{\bullet}{\gamma}\right) = 0.$$

- *Simplified Phan-Thien–Tanner model*: $N_2\left(\overset{\bullet}{\gamma}\right) = 0$.

$$f[tr\mathbf{S}]\mathbf{S} + \lambda\overset{\nabla}{\mathbf{S}} = 2\mu\mathbf{D},$$

$$f[tr\mathbf{S}] = 1 + \frac{\varepsilon\lambda}{\mu}tr\mathbf{S}, \qquad f[tr\mathbf{S}] = \exp\left(\frac{\varepsilon\lambda}{\mu}tr\mathbf{S}\right)$$

$$\left.\begin{aligned} \eta\left(\overset{\bullet}{\gamma}\right) &= \mu \\ \eta\left(\overset{\bullet}{\gamma}\right) &= \frac{\mu}{\left(1 + \lambda^2\left(\overset{\bullet}{\gamma}\right)^2\right)^{(1-n)/2}} \end{aligned}\right\} N_1\left(\overset{\bullet}{\gamma}\right) = \frac{2\lambda\mu\overset{\bullet}{\gamma}}{f[tr\mathbf{S}]}, \qquad N_2\left(\overset{\bullet}{\gamma}\right) = 0.$$

- *Modified Phan-Thien–Tanner model*: constant $\eta\left(\overset{\bullet}{\gamma}\right)$ and $\Psi_2\left(\overset{\bullet}{\gamma}\right)$

$$f[tr\mathbf{S}]\mathbf{S} + \lambda\overset{\circ}{\mathbf{S}} = 2\eta\mathbf{D}, \qquad \mathbf{T} = -p\mathbf{1} + \mathbf{S},$$

$$\overset{\circ}{\mathbf{S}} = \overset{\bullet}{\mathbf{S}} - (\nabla\mathbf{u}^\mathrm{T} - \xi\mathbf{D})\mathbf{S} - \mathbf{S}(\nabla\mathbf{u}^\mathrm{T} - \xi\mathbf{D})^\mathrm{T},$$

$$\overset{\circ}{\mathbf{S}} = \overset{\bullet}{\mathbf{S}} - \zeta^T\mathbf{S} - \mathbf{S}\zeta - \alpha(\mathbf{DS} + \mathbf{SD}), \quad \alpha = 1 - \xi.$$

Here ζ and ξ represent the vorticity tensor and the slippage coefficient, respectively. All other symbols are as defined before in sect. 3.6.1.

Secondary flows; *Gordon–Schowalter convected derivative*: $\alpha \neq 1,\ 0,\ -1$; non-affine motion
No secondary flows; *Upper convected Oldroyd derivative*: $\alpha = 1$; affine motion
No secondary flows; *Corotational Jaumann derivative*: $\alpha = 0$; affine motion

$$f[tr\mathbf{S}] = 1 : \begin{cases} \eta\left(\dot{\gamma}\right) = \mu \\ \Psi_1\left(\dot{\gamma}\right) = 2\lambda\mu \\ \Psi_2\left(\dot{\gamma}\right) = -\xi\lambda\mu \end{cases}$$

- *Fluid of order two or equivalently Rivlin–Ericksen fluid of grade 2*:

$$S_{kl} = \mu\,\mathrm{A}_{kl}^{(1)} + \frac{1}{4}\alpha_1 \mathrm{A}_{km}^{(1)}\mathrm{A}_{ml}^{(1)} + \alpha_2 \mathrm{A}_{kl}^{(2)}$$

Here $\mathbf{A}^{(1)}$ and $\mathbf{A}^{(2)}$ are the first and the second Rivlin-Ericksen tensors.
No secondary flows: constant μ
Secondary flows: replace μ with a point function in the field; apparent viscosity $\eta = f(tr\mathbf{D}^2)$.

- Corotational *Maxwell fluid*:

$S_{kl} + \lambda\overset{\circ}{S}kl = 2\mu D_{kl} \rightarrow$ *No secondary flows*:

$$\left\{ \eta\left(\dot{\gamma}\right) = \frac{\mu}{1 + \lambda^2\dot{\gamma}} \; ; \; \Psi_2\left(\dot{\gamma}\right) = -\frac{\lambda\mu}{1 + \lambda^2\dot{\gamma}} \right\} : \Psi_2\left(\dot{\gamma}\right) = -\lambda\eta\left(\dot{\gamma}\right)$$

Secondary flows: Replace constant μ with a point function; the apparent viscosity is given as $\eta = f(tr\mathbf{D}^2)$

$$\eta = \left\{ \begin{array}{ll} \mu & \text{when } \mathrm{II}_{\mathbf{D}} \leq \mathrm{II}_{\mathbf{D}}^0 \\ \mu\left(\mathrm{II}_{\mathbf{D}}/\mathrm{II}_{\mathbf{D}}^0\right)^{m-1} & \text{when } \mathrm{II}_{\mathbf{D}} > \mathrm{II}_{\mathbf{D}}^0 \end{array} \right\}, \quad \mathrm{II}_{\mathbf{D}} = \left(D_{kl}D_{lk}\right)^{1/2}$$

Appendix B
Turbulence Closure Models for Linear Fluids and Secondary Flows

A survey of turbulence closure models can be found in Mellor and Herring [254]. Two examples of second-order closure models for homogeneous turbulent flow of linear fluids are collected in this Appendix together with an example of near-wall turbulence closure model. In keeping with the tradition in the literature, the Reynolds stress tensor is denoted by $\boldsymbol{\tau}$.

In the *Launder, Reece, and Rodi* [255] model, the following closure relations for the higher-order turbulence correlations are postulated:

$$
\begin{aligned}
\overline{u_k' u_l' u_m'} &= -\frac{B_1}{\rho^2}\frac{k}{\varepsilon}\left(\tau_{mn}\frac{\partial \tau_{kl}}{\partial x_n} + \tau_{ln}\frac{\partial \tau_{mk}}{\partial x_n} + \tau_{kn}\frac{\partial \tau_{ml}}{\partial x_n}\right), \qquad \overline{p' u_k'} = 0 \\
\Pi_{ij} &= \overline{p'\left(\frac{\partial u_k'}{\partial x_l} + \frac{\partial u_l'}{\partial x_k}\right)} = B_2\frac{\varepsilon}{k}\left(\tau_{kl} - \frac{1}{3}\tau_{mm}\delta_{kl}\right) \\
&\quad - \frac{(B_3+8)}{11}\left(\tau_{km}\frac{\partial \bar{u}_l}{\partial x_m} + \tau_{lm}\frac{\partial \bar{u}_k}{\partial x_m} - \frac{2}{3}\tau_{mn}\frac{\partial \bar{u}_m}{\partial x_n}\delta_{kl}\right) \\
&\quad - \frac{(8B_3-2)}{11}\left(\tau_{km}\frac{\partial \bar{u}_m}{\partial x_l} + \tau_{lm}\frac{\partial \bar{u}_m}{\partial x_k} - \frac{2}{3}\tau_{mn}\frac{\partial \bar{u}_m}{\partial x_n}\delta_{kl}\right) \\
&\quad - \frac{(30B_3-2)}{55}\rho k\left(\frac{\partial \bar{u}_k}{\partial x_l} + \frac{\partial \bar{u}_l}{\partial x_k}\right) \\
&\nu\,\overline{\frac{\partial u_k'}{\partial x_m}\frac{\partial u_l'}{\partial x_m}} = \frac{1}{3}\varepsilon\delta_{kl}
\end{aligned}
\tag{B.1}
$$

The scalar dissipation rate ε is determined from its transport equation, Launder et al. [255],

D.A. Siginer, *Developments in the Flow of Complex Fluids in Tubes*,
DOI 10.1007/978-3-319-02426-4

$$\frac{\mathrm{D}\varepsilon}{\mathrm{D}t} = B_4 \frac{\varepsilon}{\rho K} \tau_{kl} \frac{\partial \bar{u}_k}{\partial x_l} - B_5 \frac{\varepsilon^2}{K} + B_6 \frac{\partial}{\partial x_k} \left(\frac{k}{\rho \varepsilon} \tau_{kl} \frac{\partial \varepsilon}{\partial x_l} \right)$$

$B_i, i = 1, \ldots, 6$ are dimensionless constants. It can be shown that this model leads to a non-zero normal Reynolds stress difference $(\tau_{yy} - \tau_{xx})$ unless the constants emanating from third-order diffusion correlation and the pressure–strain correlation terms are such that $B_1 = 0$, $\frac{8B_3-2}{11} = 0$. Thus, the origin of secondary flows in this model lies with third-order diffusion correlation and the pressure–strain correlation terms.

In the *Rotta–Kolmogorov* second-order closure model, Rotta [410–413], Kolmogorov [267], the following closure relations are assumed to be valid, Mellor and Herring [254],

$$\overline{u'_k u'_l u'_m} = C_1 \frac{\sqrt{k} l}{\rho} \left(\frac{\partial \tau_{kl}}{\partial x_m} + \frac{\partial \tau_{ml}}{\partial x_k} + \frac{\partial \tau_{mk}}{\partial x_l} \right)$$

$$\overline{p' u'_k} = -C_2 \rho \sqrt{k} l \frac{\partial k}{\partial x_k}$$

$$\nu \overline{\frac{\partial u'_k}{\partial x_m} \frac{\partial u'_l}{\partial x_m}} = \frac{1}{3} \frac{k^{3/2}}{l} \delta_{kl}$$

$$\overline{p' \left(\frac{\partial u'_k}{\partial x_l} + \frac{\partial u'_l}{\partial x_k} \right)} = C_3 \frac{\sqrt{k}}{l} \left(\tau_{kl} - \frac{1}{3} \tau_{mm} \delta_{kl} \right) + C_4 \rho k \left(\frac{\partial \bar{u}_k}{\partial x_l} + \frac{\partial \bar{u}_l}{\partial x_k} \right)$$

The transport equation for the length scale l assumes the following form, Mellor and Herring [244],

$$\frac{\mathrm{D}}{\mathrm{D}t}(kl) = \frac{\partial}{\partial x_k} \left[\left(\nu + C_5 l \sqrt{k} \right) \frac{\partial}{\partial x_k}(kl) + C_6 l k^{3/2} \frac{\partial l}{\partial x_k} \right] + C_7 \frac{l}{\rho} \tau_{kl} \frac{\partial \bar{u}_k}{\partial x_l} - C_8 k^{3/2}$$

$C_i, i = 1, \ldots, 8$ are dimensionless constants to be determined from experiments. It is straightforward to show that the *Rotta–Kolmogorov* second-order closure model gives rise to secondary flows as the Reynolds stress difference $(\tau_{yy} - \tau_{xx})$ is non-zero unless the constants C_1 and C_2 emanating from third-order diffusion correlation and the pressure–strain correlation terms are zero. It should be noted that in both cases the source of secondary flows lies with the third-order correlation terms. Turbulence calculations of secondary flow in pipes of non-circular cross section using second-order closure models are difficult as the problem is three dimensional, and the solution requires simultaneous solution of 11 coupled non-linear partial differential equations for the 11 unknowns $\bar{\mathbf{u}}$, $\boldsymbol{\tau}$, l and the stream function ψ.

In the near-wall *Lai and So* [265] model, the Reynolds stress closure is based on Eq. (6.8). Numerous models have been proposed for the third-order diffusion correlation $\mathcal{D}^T_{ij}$. Among these the four most popular models are due to Daly and Harlow [414], Hanjalic and Launder [415], Shir [416], and Cormack et al. [417]. *Lai and So* [255] adopt the evolution equation for $\mathcal{D}_{ij}^T$ proposed by Hanjalic and Launder [415] because it gives the best results among all four for the test case of backward-facing step flow and further that the contribution of $\mathcal{D}_{ij}^T$ to the Reynolds stress equations is small in the near-wall region as compared to $\nu\Delta\tau_{ij}$.

$$\mathcal{D}_{ij}^T = \frac{\partial \overline{u_i' u_j' u_k'}}{\partial x_k} = \frac{\partial}{\partial x_k}\left\{C_s \frac{k}{\varepsilon}\left[\tau_{il}\frac{\partial \tau_{jk}}{\partial x_l} + \tau_{jl}\frac{\partial \tau_{ki}}{\partial x_l} + \tau_{kl}\frac{\partial \tau_{ij}}{\partial x_l}\right]\right\}$$

where C_s is a model constant, and $k = \frac{1}{2}\tau_{ii}$ and $\varepsilon = \nu\overline{\left(\frac{\partial u_i'}{\partial x_k}\right)^2}$. A model for ε_{ij} is proposed along the suggestions of Hanjalic and Launder [418] and Launder and Reynolds [419].

$$\varepsilon_{ij} = \frac{2}{3}\varepsilon\left(1 - f_{w,1}\right)\delta_{ij} + f_{w,1}\frac{\varepsilon}{k}\frac{\tau_{ij} + \tau_{ik}n_k n_j + \tau_{jk}n_k n_i + n_i n_j \tau_{kl} n_k n_l}{1 + \frac{3\tau_{kl}n_l n_k}{2k}}$$

$$f_{w,1} = \exp\left[-\left(\frac{k^2}{150\nu\varepsilon}\right)^2\right]$$

where $n_k = (0,1,0)$ is the unit vector normal to the wall. The model (Eq. B.1) for *the pressure redistribution* Π_{ij} of Launder et al. [255] is adopted for Π_{ij} as it has been determined that it is quite successful for a wide variety of complex flows, and a model for $\mathcal{P}_{ij}$ is proposed,

$$\mathcal{P}_{ij} = f_{w,1}\left\{C_1\frac{\varepsilon}{k}\left(\tau_{ij} - \frac{2}{3}\delta_{ij}k\right) - \frac{\varepsilon}{k}\left(\tau_{ik}n_k n_j + \tau_{jk}n_k n_i\right) + \alpha\left(P_{ij} - \frac{1}{3}\delta_{ij}P_{ii}\right)\right\}$$

$$P_{ij} = -\tau_{ik}\frac{\partial \overline{u}_j}{\partial x_k} - \tau_{jk}\frac{\partial \overline{u}_i}{\partial x_k}$$

where α is a model constant. Using these equations and the Reynolds stress transport equations (Eq. 6.8), Lai and So [255] build an asymptotically correct k-ε closure of turbulence for the near wall:

$$\frac{\mathrm{D}k}{\mathrm{D}t} = \frac{\partial}{\partial x_j}\left(\nu\frac{\partial k}{\partial x_j}\right) + \frac{\partial}{\partial x_j}\left(\frac{\nu_t}{\sigma_k}\frac{\partial k}{\partial x_j}\right) + \frac{1}{2}P_{ii} - \varepsilon - f_{w,1}\frac{\varepsilon}{k}\overline{u_2' u_2'}$$

$$\frac{\mathrm{D}\varepsilon}{\mathrm{D}t} = \frac{\partial}{\partial x_j}\left(\nu\frac{\partial \varepsilon}{\partial x_j}\right) + \frac{\partial}{\partial x_j}\left(\frac{\nu_t}{\sigma_\varepsilon}\frac{\partial \varepsilon}{\partial x_j}\right) + \frac{1}{2}C_{\varepsilon 1}\left(1 + \sigma f_{w,2}\right)\frac{\varepsilon}{k}P_{ii} - C_{\varepsilon 2}f_\varepsilon\frac{\varepsilon\tilde{\varepsilon}}{k} + \xi$$

$$-\tau_{ij} = \nu_t\left(\frac{\partial \overline{u}_i}{\partial x_j} + \frac{\partial \overline{u}_j}{\partial x_i}\right) - \frac{2}{3}\delta_{ij}k$$

$$\nu_t = C_\mu f_\mu \frac{k^2}{\varepsilon}, \qquad f_\mu = 1 - \exp\left(\frac{C_3 y u_\tau}{\nu}\right)$$

$$f_\varepsilon = 1 - \frac{2}{9}\exp\left(-\left(\frac{1}{6}\frac{k^2}{\nu\varepsilon}\right)^2\right), \qquad \widetilde{\varepsilon} = \varepsilon - 2\nu\left(\frac{\partial\sqrt{k}}{\partial y}\right)^2$$

$$\xi = f_{w,2}\left[\left(\frac{7}{9}C_{\varepsilon 2} - 2\right)\right]\frac{\varepsilon\overline{\varepsilon}}{k} - \frac{1}{2}\frac{\overline{\varepsilon}^2}{k}$$

$$\overline{\varepsilon} = \varepsilon - \nu\frac{\partial^2 k}{\partial x_j \partial x_j}, \qquad f_{w,2} = \exp\left[-\left(\frac{k^2}{64\varepsilon\nu}\right)^2\right]$$

The functions $f_{w,i}\quad i = 1, 2$ guarantee that the corresponding expressions will asymptote to the Kolmogorov's model far away from the wall. σ_k, σ_ε, $C_{\varepsilon i}\quad i = 1, 2,$ and C_3 as well as C_μ are model constants.

References

1. Shah RK, London AL (1978) Laminar flow forced convection in ducts, Advances in heat transfer series. Academic Press, New York, NY
2. Shah RK, Bhatti MS (1987) Laminar convective heat transfer in ducts. In: Kakac S, Shah RK, Aung W (eds) Handbook of convective single phase heat transfer. Wiley-Interscience, New York, NY
3. Eckert ERG, Irvine TF Jr (1960) Pressure drop and heat transfer in a duct with triangular cross-section. J Heat Tran 82:125–138
4. Carlson LW, Irvine TF Jr (1961) Fully developed pressure drop in triangular shaped ducts. J Heat Tran 83:441–444
5. Natarajan NM, Lakshmanan SM (1972) Laminar flow in rectangular ducts: prediction of velocity profiles and friction factor. Indian J Tech 10:435–461
6. Schecter RS (1961) On the steady flow of a non-Newtonian fluid in cylinder ducts. AIChE J 7:445–457
7. Wheeler JA, Wissler EH (1966) The friction factor-reynolds number relation for the steady flow of pseudoplastic fluids through rectangular ducts. AIChE J 11:207–221
8. Metzner AB, Reed JC (1955) Flow of non-Newtonian fluids – correlation of the laminar, transition and turbulent flow regions. AIChE J 1:434–449
9. Metzner AB (1965) Heat transfer in non-Newtonian fluids. Adv Heat Tran 2:357–397
10. Cho YI, Hartnett JP (1982) Non-Newtonian fluids in circular pipe flow. Adv Heat Tran 15:59–141
11. Kozicki W, Chou CH, Tiu C (1966) Non-Newtonian flow in ducts of arbitrary cross-sectional shape. Chem Eng Sci 21(8):665–679
12. Hartnett JP, Kwack EY, Rao BK (1986) Hydrodynamic behavior of non-Newtonian fluids in a square duct. J Rheol 30(S):545–559
13. Hartnett JP, Kostic M (1985) Heat transfer to a viscoelastic fluid in laminar flow through a rectangular channel. Int J Heat Mass Tran 28(6):1147–1155
14. Bhamidipaty KR (1988) Heat transfer to viscoelastic fluids in a 5:1 rectangular duct. PhD dissertation, University of Illinois at Chicago
15. Siginer DA, Letelier M (2011) Laminar flow of non-linear viscoelastic fluids in straight tubes of arbitrary contour. Int J Heat Mass Tran 54(9–10):2188–2202
16. Johnson MW Jr, Segalman D (1977) Model for viscoelastic fluid behavior which allows non-affine deformation. J Non-Newton Fluid Mech 2(3):255–270
17. Phan-Thien N (1978) Non-linear network viscoelastic model. J Rheol 22(3):259–283
18. Phan-Thien N, Tanner RI (1977) New constitutive equation derived from network theory. J Non-Newton Fluid Mech 2(4):353–365

D.A. Siginer, *Developments in the Flow of Complex Fluids in Tubes*,
DOI 10.1007/978-3-319-02426-4

19. Chai MS, Yeow TL (1990) Modelling of fluid M1 using multiple-relaxation-time constitutive equations. J Non-Newton Fluid Mech 35(2–3):459–470
20. Ahmeda M, Normandin M, Clermont JR (1995) Calculation of fully developed flows of complex fluids in pipes of arbitrary shape using a mapped circular domain. Comm Numer Meth Eng 11:813–820
21. Papanastasiou AC, Scriven LE, Macosko CW (1983) Integral constitutive equation for mixed flows: viscoelastic characterization. J Rheol 27(4):387–410
22. Bhatti MS, Shah RK (1987) Turbulent and transition flow convective heat transfer in ducts. In: Kakac S, Shah RK, Aung W (eds) Handbook of convective single phase heat transfer. Wiley-Interscience, New York, NY
23. Hartnett JP, Koh JC, McComas ST (1962) A comparison of predicted and measured friction factor for turbulent flow through rectangular ducts. J Heat Tran 84:82–87
24. Jones OC Jr (1981) An improvement in the calculation of turbulent friction in rectangular ducts. J Fluid Eng 98:173–178
25. Dodge DW, Metzner AB (1959) Turbulent flow of non-Newtonian fluids. AIChE J 5:189–204
26. Kozicki W, Chou CH, Tiu C (1966) Non-Newtonian flow in ducts of arbitrary cross-sectional shape. Chem Eng Sci 21:665–679
27. Kostic M, Hartnett JP (1984) Predicting turbulent friction factors of non-Newtonian fluids in non-circular ducts. Int Comm Heat Mass Tran 11:345–352
28. Hartnett JP, Rao BK (1987) Heat transfer and pressure drop for purely viscous non-Newtonian fluids in turbulent flow through rectangular passages. Warme Stoffubertrag 21:261–267
29. Virk PS, Merrill EW, Mickley HS, Smith KA, Mollo-Christensen EL (1967) The Toms phenomenon: turbulent pipe flow of dilute polymer solutions. J Fluid Mech 30:305–325
30. Virk PS, Mickley HS, Smith KA (1970) The ultimate asymptote and mean flow structure in Toms' phenomenon. J Appl Mech 37:488–493
31. Virk PS (1971) An elastic sublayer model for drag reduction by dilute polymer solutions. J Fluid Mech 45:417–430
32. Virk PS (1971) Drag reduction in rough pipes. J Fluid Mech 45:225–246
33. Hartnett JP, Kwack E (1985) Empirical correlations of turbulent friction factors and heat transfer coefficients of aqueous polyacrylamide solutions. Proceedings, Int. Symp. Heat Transfer, 1985
34. Ghajar AJ, Azar MY (1988) Empirical correlations for friction factor in drag-reducing turbulent pipe flows. Int Comm Heat Mass Tran 15(6):705–718
35. Schroeder CB, Jeffrey KR (2011) Rheo-nmr of the secondary flow of non-Newtonian fluids in square ducts. Phys Rev Lett 106(4):046001
36. Callaghan PT (1991) Principles of nuclear magnetic resonance microscopy. Oxford Univ Press, Oxford
37. Toms BA (1948) Some observations on the flow of linear polymer solutions through straight tubes at large Reynolds numbers. Proc First Int. Cong. Rheol., North Holland, Amsterdam, 2, pp. 135–141
38. Mysels KJ (1949) U.S. Patent 2, 492: 173
39. Mysels KJ (1971) Early experiences with viscous drag reduction. AIChE Chem Eng Prog Symp Ser III 67:45–49
40. Forrest F, Grierson GA (1931) Friction losses in cast iron pipe carrying paper stock. Paper Trade J 92(22):39–41
41. Kubo T, Ogata S (2012) Flow properties of bamboo fiber suspensions. ASME 2012 international mechanical engineering congress & exposition. Paper IMECE2012-87283, Houston, TX, November 9–15, 2012
42. Burger ED, Munk WR, Wahl HA (1982) Flow increase in the trans Alaska pipeline through use of a polymeric drag-reducing additive. J Petrol Tech 34(2):377–386

43. Kline SJ, Reynolds WC, Schraub FA, Runstadler PW (1967) The structure of turbulent boundary layers. J Fluid Mech 30(4):741–773
44. Corino ER, Brodkey RS (1969) A visual investigation of the wall region in turbulent flow. J Fluid Mech 37(1):1–30
45. Sirkar KK, Hanratty TJ (1970) The limiting behaviour of the turbulent transverse velocity component close to a wall. J Fluid Mech 44(3):605–614
46. Kim HT, Kline SJ, Reynolds WC (1971) The production of turbulence near a smooth wall in a turbulent boundary layer. J Fluid Mech 50(1):133–160
47. Eckelman LD, Fortuna G, Hanratty TJ (1972) Drag reduction and the wavelength of flow-oriented wall eddies. Nat Phys Sci 236:94–96
48. Donohue GL, Tiederman WG, Reischman MM (1972) Flow visualization in the near-wall region in a drag reducing flow. J Fluid Mech 56(3):559–575
49. Fortuna G, Hanratty TJ (1972) The influence of drag-reducing polymers on turbulence in the viscous sublayer. J Fluid Mech 53(3):575–586
50. Rudd MJ (1972) Velocity measurements made with a laser Doppler meter on the turbulent pipe flow of a dilute polymer solution. J Fluid Mech 51(4):673–685
51. Oldaker DK, Tiederman WG (1977) Spatial structure of the viscous sublayer in drag-reducing channel flows. Phys Fluids 20(10):S133
52. Achia BU, Thompson DW (1977) Structure of the turbulent boundary layer in drag reducing pipe flow. J Fluid Mech 81(3):439–464
53. Lumley JL (1977) Drag reduction in two phase and polymer flows. Phys Fluids 20(10):S64
54. Reischman MM, Tiederman WG (1975) Laser-doppler anemometer measurements in drag-reducing channel flows. J Fluid Mech 70(2):369–392
55. Tiederman WG, Luchik TS, Boggard DG (1985) Wall-layer structure and drag reduction. J Fluid Mech 156:419–437
56. McComb WD, Rabie LH (1982) Local drag reduction due to injection of polymer solutions into turbulent flow in a pipe. Part I: dependence on local polymer concentration. AIChE J 28 (4):547–557
57. McComb WD, Rabie LH (1982) Local drag reduction due to injection of polymer solutions into turbulent flow in a pipe. Part II: laser-Doppler measurements of turbulent structure. AIChE J 28(4):558–565
58. Usui H, Kodama M, Sano Y (1988) Laser-Doppler measurements of turbulence structure in a drag-reducing pipe flow with polymer injection. J Chem Eng Jpn 21(2):134–140
59. Luchik TS, Tiederman WG (1988) Turbulent structure in low concentration drag-reducing channel flows. J Fluid Mech 190:241–263
60. Walker DT, Tiederman WG (1990) Turbulent structure in a channel flow with polymer injection at the wall. J Fluid Mech 218:377–403
61. Metzner AB, Park MG (1964) Turbulent flow characteristics of viscoelastic fluids. J Fluid Mech 20(2):91–303
62. Seyer FA, Metzner AB (1969) Turbulence phenomena in drag-reducing systems. AIChE J 15:426–434
63. Metzner AB (1977) Polymer solution and fiber suspension rheology and their relationship to turbulent drag reduction. Phys Fluids 20:S145
64. Bewersdorff HW, Berman NS (1988) The influence of flow-induced non-Newtonian fluid properties on turbulent drag reduction. Rheol Acta 27(2):130–136
65. Graham MD (2004) Drag reduction in turbulent flow of polymer solutions. In: Rheology reviews. The British Society of Rheology, London, pp 143–170
66. Wang Y, Yu B, Zakin JL, Shi H (2011) Review on drag reduction and its heat transfer by additives. Adv Mech Eng. Article ID 478749, doi: 10.1155/2011/478749
67. Lumley JL (1969) Drag reduction by additives. Annu Rev Fluid Mech 1:367–384
68. Lumley JL (1973) Drag reduction in turbulent flow by polymer additives. J Polymer Sci Macromol Rev 7:263–290
69. Schowalter WR (1978) Mechanics of non-Newtonian fluids. Pergamon Press, London

70. Kohn MC (1974) Criteria for the onset of drag reduction. AIChE J 20(1):185–188
71. den Toonder JMJ, Nieuwstadt FTM, Kuiken GDC (1995) Role of elongational viscosity in the mechanism of drag reduction by polymer additives. Appl Sci Res 54(2):95–123
72. Orlandi P (1995) A tentative approach to the direct simulation of drag reduction by polymers. J Nonnewton Fluid Mech 60(2–3):277–301
73. Siginer DA (2014) Stability of non-linear constitutive formulations for viscoelastic fluids. Springer, New York, NY
74. Sureshkumar R, Beris AN, Handler RA (1997) Direct numerical simulation of the turbulent channel flow of a polymer solution. Phys Fluid 9(3):743–757
75. Leal LG (1990) Dynamics of dilute polymer solutions. In: Gyr A (ed) Structure of turbulence and drag reduction. Springer, New York, NY, pp 155–185
76. Hershey HC, Zakin JL (1967) A molecular approach to predicting the onset of turbulent drag reduction in the turbulent flow of dilute polymer solutions. Chem Eng Sci 22(12):1847–1857
77. Virk PS, Wagger DL (1989) Aspects of mechanism in type B drag reduction. In Proceedings of the 2nd IUTAM Symp. on structure of turbulence and drag reduction, Zurich, Switzerland, 1989
78. Virk PS (1975) Drag reduction fundamentals. AIChE J 21(4):625–656
79. Hoyt JW (1986) Drag reduction. In: Mark HF et al (eds) Encyclopedia of polymer science and engineering, vol 5, 2nd edn. John Wiley & Sons, New York, NY, pp 129–151
80. Berman NS (1978) Drag reduction by polymers. Annu Rev Fluid Mech 10:47–64
81. Fischer P (2000) Time dependent flow in equimolar micellar solutions: transient behaviour of the shear stress and first normal stress difference in shear induced structures coupled with flow instabilities. Rheol Acta 39(3):234–240
82. Ohlendorf D, Interthal W, Hoffmann H (1986) Surfactant systems for drag reduction: physico-chemical properties and rheological behaviour. Rheol Acta 25(5):468–486
83. Gasljevic K, Matthys EF, Aguilar G (2001) On two distinct types of drag-reducing fluids, diameter scaling, and turbulent profiles. J Non-Newton Fluid Mech 96(3):405–425
84. Aguilar G, Gasljevic K, Matthys EF (1999) Coupling between heat and momentum transfer mechanisms for drag-reducing polymer and surfactant solutions. J Heat Tran 121(4):796–802
85. White A, Hemmings JAG (1976) Drag reduction by additives—Review and bibliography. BHRA Fluid Engineering, Cranfield, UK
86. McComb WD (1990) The physics of fluid turbulence. Oxford University Press, New York, NY
87. Gyr A, Bewersdorff H-W (1989) Change of structures close to the wall of a turbulent flow in drag reducing fluids. In: Gyr A (ed) Structure of turbulence and drag reduction. Springer, New York, NY, pp 215–222
88. Gyr A, Bewersdorff H-W (1995) Drag reduction of turbulent flows. Kluwer, Dordrecht
89. White CM, Mungal MG (2008) Mechanics and prediction of turbulent drag reduction with polymer additives. Annu Rev Fluid Mech 40:235–256
90. Manfield PD, Lawrence CJ, Hewitt GF (1999) Drag reduction with additives in multiphase flow: a literature survey. Multiphas Sci Tech 11(3):197–221
91. Nadolink RH, Haigh WW (1995) Bibliography on skin friction reduction with polymers and other boundary-layer additives. Appl Mech Rev 48:351–460
92. Ge W (2008) Studies on the nanostructure, rheology and drag reduction characteristics of drag reducing cationic surfactant solutions. PhD thesis, The Ohio State University
93. Mollica F, Rajagopal KR (1997) Secondary deformations due to axial shear of the annular region between two eccentrically placed cylinders. J Elasticity 48(2):103–123
94. Mollica F, Rajagopal KR (1999) Secondary flows due to axial shearing of a third grade fluid between two eccentrically placed cylinders. Int J Eng Sci 37:411–429
95. Rivlin RS (1957) The relation between the flow of non-Newtonian fluids and turbulent Newtonian fluids. Q Appl Math 15:212–214
96. Speziale CG (1983) Closure models for rotating two dimensional turbulence. Geophys Astrophys Fluid Dyn 23:69–80

97. Speziale CG (1982) On turbulent secondary flows in pipes of non-circular cross section. Int J Eng Sci 7:863–872
98. Huang YN, Rajagopal KR (1995) On necessary and sufficient conditions for turbulent secondary flows in a straight tube. Mat Model Meth Appl Sci 5:111–123
99. Nikuradse J (1930) Turbulente Stromung in Nicht-Kreisformigen Rohren. Ingenieur Archiv 1:306–332
100. Emery AF, Neighbors PK, Gessner FB (1980) The numerical prediction of developing turbulent flow and heat transfer in a square duct. J Heat Tran 102:51–57
101. Ericksen JL (1956) Overdetermination of the speed in rectilinear motion of non-Newtonian fluids. Q Appl Math 14:318–321
102. Green AE, Rivlin RS (1956) Steady flow of non-Newtonian fluids through tubes. Q Appl Math 14:299–308
103. Langlois WE, Rivlin RS (1963) Slow steady – state flow of viscoelastic fluids through non-circular tubes. Rend Math 22:169–185
104. Rivlin RS (1964) Second order effects in elasticity, plasticity and fluid dynamics. Pergamon, London
105. Pipkin AC (1963) Proc. 4th Int. congress rheology, vol 1. Interscience, New York, NY, p 213
106. Wheeler JA, Wissler EH (1966) Steady flow of non-Newtonian fluids in a square duct. Tran Soc Rheol 10(1):353–367
107. Giesekus H (1965) Sekundärströmungen in viscoelastischen Flüssigkeiten bei stationärer und periodischer Bewegung. Rheol Acta 4(2):85–101
108. Semjonow V (1967) Sekundärströmungen hochpolymerer Schmelzen in einem Rohr von elliptischem Querschnitt. Rheol Acta 6(2):171–173
109. Fosdick RL, Kao BG (1978) Transverse deformations associated with rectilinear shear in elastic solids. J Elasticity 8:117–142
110. Oldroyd JG (1965) Some steady flows of the general elastico-viscous liquid. Proc Roy Soc Lond A 283(1392):115–133
111. Fosdick RL, Serrin J (1973) Rectilinear steady flow of simple fluids. Proc Roy Soc Lond A 332(1590):311–333
112. Stone DE (1957) On the non-existence of rectilinear motion in plastic solids and non-Newtonian fluids. Q Appl Math 15:257–262
113. Pipkin AC, Rivlin RS (1963) Normal stresses in flow through tubes of non-circular cross-section. ZAMP 14:738–745
114. Criminale WO Jr, Ericksen JL, Filbey GL (1957) Steady shear flow of non-Newtonian fluids. Arch Rat Mech Anal 1(1):410–417
115. Truesdell C, Noll W (1992) The non-linear field theories of mechanics, 2nd edn. Springer, Berlin
116. Dunn JE, Rajagopal KR (1995) Fluids of differential type: critical review and thermodynamic analysis. Int J Eng Sci 33(5):689–769
117. Rivlin RS, Ericksen JL (1955) Stress deformation relations for isotropic materials. J Rat Mech Anal 4(2):323–425 (New title: *Indiana Univ. Math. J.*)
118. Coleman BD, Noll W (1960) An approximation theorem for functionals with applications in continuum mechanics. Arch Rat Mech Anal 6(1):355–370
119. Fosdick RL, Rajagopal KR (1980) Thermodynamics and stability of fluids of third grade. Proc Roy Soc Lond A 369(1738):351–377
120. Müller I, Wilmanski K (1986) Extended thermodynamics of a non-Newtonian fluid. Rheol Acta 25:335–349
121. Dunn JR, Fosdick RL (1974) Thermodynamics, stability and boundedness of fluids of complexity 2 and fluids of second grade. Arch Rat Mech Anal 56(3):191–252
122. Joseph DD (1981) Instability of the rest state of fluids of arbitrary grade greater than one. Arch Rat Mech Anal 75(3):251–256
123. Müller I (1985) Thermodynamics. Pitman Publishing, London

124. Lebon G, Cloot A (1988) An extended thermodynamic approach to non-newtonian fluids and related results in Marangoni instability problem. J Non-Newton Fluid Mech 28(1):61–76
125. Depireux N, Lebon G (2001) An extended thermodynamics modeling of non-Fickian diffusion. J Non-Newton Fluid Mech 96(1–2):105–117
126. Jones JR (1964) Secondary flow of non-Newtonian fluids between eccentric cylinders in relative motion. ZAMP 15:329–341
127. Dodson AG, Townsend P, Walters K (1974) Non-Newtonian flow in pipes of non-circular cross section. Comput Fluids 2:317–338
128. Townsend P, Walters K, Waterhouse WM (1976) Secondary flows in pipes of square cross-section and the measurement of the second normal stress difference. J Non-Newton Fluid Mech 1(2):107–123
129. McLeod JB (1972) Over-determined systems and the rectilinear steady flow of simple fluids. In: Sleeman BD, Michael ID (eds) Proc conference on ordinary and partial differential equations, vol 415, Lecture notes in mathematics. Springer, Berlin, pp 193–204
130. Speziale CG (1984) On the development of non-Newtonian secondary flows in tubes of non-circular cross-section. Acta Mech 51:85–95
131. Huang YN, Rajagopal KR (1994) On necessary conditions for the secondary flow of non-Newtonian fluids in straight tubes. Int J Eng Sci 32(8):1277–1281
132. Yue P, Dooley J, Feng JJ (2008) A general criterion for viscoelastic secondary flow in pipes of noncircular cross section. J Rheol 52(1):315–332
133. Speziale CG, Thangam S. Notes on secondary flows of non-newtonian fluids, unpublished notes, personal communication
134. Thangam S, Speziale CG (1987) Non-Newtonian secondary flows in ducts of rectangular cross-section. Acta Mech 68:121–138
135. Speziale CG, Thangam S (1986) Instabilities with reduced frictional drag in the rotating channel flow of dilute polymer solutions. Acta Mech 60:17–40
136. Hart JE (1971) Instability and secondary motion in a rotating channel flow. J Fluid Mech 45:341–351
137. Lezius DK, Johnston JP (1976) Roll-cell Instabilities in rotating laminar and turbulent channel flows. J Fluid Mech 77:153–174
138. Speziale CG (1982) Numerical study of viscous flow in rotating rectangular ducts. J Fluid Mech 122:251–271
139. Speziale CG, Thangam S (1983) Numerical study of secondary flows and roll-cell instabilities in rotating channel flow. J Fluid Mech 130:377–395
140. Speziale CG (1983) On the non-linear stability of rotating Newtonian and non-Newtonian fluids. Acta Mech 49:263–273
141. Denn MM, Roisman JJ (1969) Rotational stability and measurement of normal stress functions in dilute polymer solutions. AIChE J 15:454–459
142. Denn MM, Sun Z-S, Rushton BD (1971) Torque reduction in finite amplitude secondary flows of dilute polymer solutions. Tran Soc Rheol 15(3):415–431
143. Ginn RF, Denn MM (1969) Rotational stability in viscoelastic liquids: theory. AIChE J 15:450–454
144. Hartnett JP, Kostic M (1989) Heat transfer to Newtonian and non-Newtonian fluids in rectangular ducts, vol 19, Advances in Heat Transfer. Academic Press, New York, NY, pp 247–356
145. Crochet MJ, Davies AR, Walters K (1984) Numerical simulation of non-Newtonian flows. Elsevier, Amsterdam
146. Keunings R (1988) Simulation of viscoelastic fluid flow. In: Tucker CL III (ed) Fundamentals of computer modeling for polymer processing. Hanser, Munich, Chapter 9
147. Gervang B, Larsen PS (1991) Secondary flows in straight ducts of rectangular cross section. J Non-Newton Fluid Mech 39(3):217–237

148. Xue SC, Phan-Thien N, Tanner RI (1995) Numerical study of secondary flows of viscoelastic fluids in straight pipes by an implicit finite volume method. J Non-Newton Fluid Mech 59 (2–3):191–213
149. Patankar SV (1980) Numerical heat transfer and fluid flow. Mc Graw Hill, New York, NY
150. Van Doormaal JP, Raithby GD (1984) Enhancements of the simple method for predicting incompressible fluid flows. Numer Heat Tran 7:147–163
151. Issa RI (1985) Solution of the implicitly discretized fluid flow equations by operator-splitting. J Comput Phys 62(1):40–65
152. Debbaut B, Avalosse T, Dooley J, Hughes K (1997) On the development of secondary motions in straight channels induced by the second normal stress difference: experiments and simulations. J Non-Newton Fluid Mech 69(2–3):255–271
153. Debbaut B, Dooley J (1999) Secondary motions in straight and tapered channels: experiments and three dimensional finite element simulation with a multimode differential viscoelastic model. J Rheol 43(6):1525–1545
154. Rajagopalan D, Armstrong RC, Brown RA (1990) Finite element methods for calculation of steady viscoelastic flows using constitutive equations with a Newtonian viscosity. J Non-Newton Fluid Mech 36:159–192
155. Avalosse T (1996) Numerical simulation of distributive mixing in 3-D flows. Macromol Symp 112:91–98
156. Siline M, Leonov AI (2001) On flows of viscoelastic liquids in long channels and dies. Int J Eng Sci 39(4):415–437
157. Leonov AI (1999) Constitutive equations for viscoelastic liquids: formulation, analysis and comparison with data. In: Siginer DA, De Kee D, Chhabra RP (eds) Advances in the flow and rheology of non-Newtonian fluids Part A, vol 8, Rheology series. Elsevier, New York, NY, pp 577–591
158. Simhambhatla M, Leonov AI (1995) On the rheological modeling of viscoelastic polymer liquids by stable constitutive equations. Rheol Acta 34(3):259–273
159. Meißner J (1971) Dehnungsverhalten von Polyäthylen-Schmelzen. Rheol Acta 10 (2):230–242
160. Hashemabadi SH, Etemad SG (2006) Effect of rounded corners on the secondary flow of viscoelastic fluids through non-circular ducts. Int J Heat Mass Tran 49:1986–1990
161. Siginer DA, Letelier FM (2010) Heat transfer asymptote in laminar flow of non-linear viscoelastic fluids in straight non-circular tubes. Int J Eng Sci 48(11):1544–1562
162. Zhang M, Shen X, Ma J, Zhang B (2007) Numerical study of the flow of oldroyd-3-constant fluids in a straight duct with square cross-section. Kr Aust Rheol J 19(2):67–73
163. Phan-Thien N, Huilgol RR (1985) On the stability of the torsional flow of a class of Oldroyd-type fluids. Rheol Acta 24(6):551–555
164. Russel TWF, Charles ME (1959) The effect of the less viscous liquid in the laminar flow of two immiscible liquids. Can J Chem Eng 37:18–34
165. Charles ME, Lilleleht LU (1965) Co-current stratified laminar flow of two immiscible liquids in a rectangular conduit. Can J Chem Eng 43:110–116
166. Blais P, Carlssen DJ, Surunchuk T, Wiles DM (1971) Bicomponent composites: preparation from incompatible polymers by corona treatment. Textil Res J 41:485–491
167. Southern JH, Ballman RL (1975) Additional observation on stratified bicomponent flow of polymer melts in a tube. J Polymer Sci 13(4):863–869
168. Khan AA, Han CD (1976) On the interface deformation in the stratified two-phase flow of viscoelastic fluids. Tran Soc Rheol 20(4):595–621
169. Dooley J, Hyun KS, Hughes K (1998) An experimental study on the effect of polymer viscoelasticity on layer rearrangement in coextruded structures. Polymer Eng Sci 38 (7):1060–1071
170. Dooley J, Rudolph L (2003) Viscous and elastic effects in polymer coextrusion. J Plast Film Sheet 19:111–121
171. Joseph DD (1990) Fluid dynamics of viscoelastic liquids. Springer, New York, NY

172. Yoo JY, Joseph DD (1985) Hyperbolicity and change of type in the flow of viscoelastic fluids through channels. J Non-Newton Fluid Mech 19(1):15–41
173. Ramaprian BR, Tu SW (1980) An experimental study of oscillatory pipe flow at transitional Reynolds numbers. J Fluid Mech 100:513–544
174. Tu SW, Ramaprian BR (1983) Fully developed periodic turbulent pipe flow in a tube. J Fluid Mech 137:31–58
175. Shemer L, Kit E (1984) An experimental investigation of the quasi-steady turbulent pulsating flow in a pipe. Phys Fluids 27:72–76
176. Shemer L, Wygnanski I, Kit E (1985) Pulsating flow in a pipe. J Fluid Mech 153:313–337
177. Sexl T (1930) Ueber den von E. G. Richardson entdeckten Annulareffect. Z Phys 61:349–362
178. Lambossy P (1952) Oscillations Forcées d'un Liquide Incompressible et Visqueux dans un Tube Rigide et Horizontal. Calcul de la Force de Frottement. Helvet Phys Acta 25:371–386
179. Womersley JR (1955) Oscillatory motion of a viscous liquid in a thin-walled elastic tube – I: the linear approximation for long waves. Phil Mag 46:199–221
180. Uchida S (1956) Pulsating viscous flow superposed on the steady laminar motion. Z Angew Math Phys 7:403–422
181. Pipkin AC (1964) Alternating flow of non-Newtonian fluids in tubes of arbitrary cross-section. Arch Rat Mech Anal 15(1):1–13
182. Etter I, Showalter WR (1965) Unsteady flow of an Oldroyd fluid in a circular tube. Tran Soc Rheol 9(2):351–369
183. Lanir Y, Rubin H (1971) Oscillatory flow of linear viscoelastic fluids in thin-walled elastico-viscous fluids. Rheol Acta 10:467–472
184. Walters K, Townsend P (1968) The flow of viscous and elastico-viscous liquids in straight pipes under a varying pressure gradient. Proc the 5th Int Cong Rheol, Kyoto 1968, 4: 471–483
185. Barnes HA, Townsend P, Walters K (1971) On pulsatile flow of non-Newtonian liquids. Rheol Acta 10:517–527
186. Edwards MF, Nellist DA, Wilkinson WR (1972) Pulsating flow of non-Newtonian fluids in pipes. Chem Eng Sci 27(3):545–553
187. Townsend P (1973) Numerical solutions of some unsteady flows of elastico-viscous liquids. Rheol Acta 12(1):13–18
188. DaVies JM, Bhumiratana S, Bird RB (1978) Elastic and inertial effects in pulsatile flow of polymeric liquids in circular tubes. J Non-Newton Fluid Mech 3(3):237–259
189. Böhme G, Nonn G (1979) Instätionare rohrströmung viscoelasticher flüssigkeiten, maβnahmen zur durchsatzsteigerung. Ingenieur Archiv 48:35–49
190. Manero O, Walters K (1980) On elastic effects in unsteady pipe flows. Rheol Acta 19 (3):277–284
191. Phan-Thien N (1978) On pulsatile flow of polymeric liquids. J Non-Newton Fluid Mech 4:167–176
192. Phan-Thien N (1980) Flow enhancement mechanisms of a pulsating flow of non-Newtonian liquids. Rheol Acta 19(3):285–290
193. Phan-Thien N (1981) On a pulsating flow of polymeric fluids: strain-dependent memory kernels. J Rheol 25(3):293–314
194. Vlastos G, Lerche D, Koch B, Samba O, Pohl M (1997) The effect of parallel combined steady and oscillatory shear flows on blood and polymer solutions. Rheol Acta 36(2):160–172
195. Huilgol RR, Phan-Thien N (1986) Recent advances in the continuum mechanics of visco-elastic liquids. Int J Eng Sci 24(2):161–261
196. Sundstrom DW, Kaufman A (1977) Pulsating flow of polymer solutions. Ind Eng Chem Proc Des Dev 16:320–325
197. Siginer DA (1991) On the pulsating pressure gradient driven flow of viscoelastic liquids. J Rheol 35(2):271–311
198. Siginer DA (1991) Oscillating flow of a simple fluid in a pipe. Int J Eng Sci 29 (12):1557–1567

199. Siginer DA (1993) Memory integral constitutive equations in periodic flows and rheometry. Int J Polym Mater 21:45–56
200. Green AE, Rivlin RS (1957) The mechanics of non-linear materials with memory, Part I. Arch Rat Mech Anal 1(1):1–21
201. Joseph DD (1976) Stability of fluid motions II. Springer, Berlin
202. Booij HC (1968) Influence of superimposed steady shear flow on the dynamic properties of non-Newtonian fluids. Rheol Acta 7(3):202–209
203. Booij HC (1970) Effect of superimposed steady shear flow on dynamic properties of polymeric fluids. PhD thesis, Leiden
204. Bernstein B (1969) Small shearing oscillations superposed on large steady shear of BKZ fluid. Int J Non-Lin Mech 4(2):183–200
205. Bernstein B, Fosdick RL (1970) On four rheological relations. Rheol Acta 9(2):186–193
206. Jones TER, Walters K (1971) The behavior of materials under combined steady and oscillatory shear. J Phys A Gen Phys 4(1):85–100
207. Zahorski S (1972) Flows with proportional stretch history. Arch Mech Stos 24:681
208. Zahorski S (1973) Motions with superposed proportional stretch histories as applied to combined steady and oscillatory flows of simple fluids. Arch Mech Stos 25:575
209. Goldstein C, Schowalter WR (1973) Non-linear effects in the unsteady flow of viscoelastic fluids. Rheol Acta 12(2):253–262
210. Manero O, Mena B (1977) An interesting effect in non-Newtonian flow in oscillating pipes. Rheol Acta 16(5):573–576
211. Manero O, Mena B, Valenzuela R (1978) Further developments in non-Newtonian flow in oscillating pipes. Rheol Acta 17(6):693–697
212. Simmons JM (1968) Dynamic modulus of polyisobutylene solutions in superposed steady shear flow. Rheol Acta 7(2):184–188
213. Tanner RI, Williams G (1971) On the orthogonal superposition of simple shearing and small strain oscillatory motions. Rheol Acta 10:528–538
214. Kazakia JY, Rivlin RS (1978) The influence of vibration on Poiseuille flow of a non-Newtonian fluid, I. Rheol Acta 17(3):210–226
215. Kazakia JY, Rivlin RS (1979) The influence of vibration on Poiseuille flow of a non-Newtonian fluid II. Rheol Acta 18(2):244–255
216. Joseph DD (1981) Instability of the rest state of fluids of arbitrary grade greater than one. Arch Ration Mech Anal 75(3):251–256
217. Renardy M (1984) On the domain space for constitutive laws in linear viscoelasticity. Arch Ration Mech Anal 85(1):21–26
218. Phan-Thien N (1980) The effects of random longitudinal vibration on pipe flow of a non-Newtonian liquid. Rheol Acta 19(5):539–547
219. Phan-Thien N (1981) The influence of random longitudinal vibration on channel and pipe flows of a slightly non-Newtonian liquid. J Appl Mech 48:661–664
220. Wong CM, Isayev AI (1989) Orthogonal superposition of small and large amplitude oscillations upon steady shear flow of polymer fluids. Rheol Acta 28(2):176–189
221. Kwon Y, Leonov AI (1993) Remarks on orthogonal superposition of small amplitude oscillations on steady shear flow. Rheol Acta 32(1):108–112
222. Böhme G, Voss R (1987) Unsteady shear flow of non-linear viscoelastic fluids with finite elements. J Non-Newton Fluid Mech 23:321–333
223. Siginer DA (1991) On some nearly viscometric flows. Rheol Acta 30(5):447–473
224. Siginer DA (1992) On the effect of boundary vibration on poiseuille flow of an elastico-viscous liquid. J Fluid Struct 6:719–748
225. Siginer DA, Valenzuela-Rendon A (1995) Unsteady non-viscometric flows in tubes driven by rotational boundary waves-I: longitudinal field. Int J Eng Mech 33(5):731–756
226. Andrienko YA, Siginer DA, Yanovsky YG (2000) Resonance behavior of viscoelastic fluids in Poiseuille flow and application to flow enhancement. Int J Non-Lin Mech 35:95–102

227. Fredrickson AG (1964) Principles and applications of rheology. Prentice-Hall, Englewood Cliffs, NJ
228. Siginer DA (1991) Anomalous steady flows in a tube. Rheol Acta 30(5):474–479
229. Siginer DA, Valenzuela-Rendon A (1993) Energy considerations in the flow enhancement of viscoelastic liquids. J Appl Mech 60(2):344–352
230. Mai T-Z, Davies AMJ (1996) Laminar pulsatile two-phase non-Newtonian flow through a pipe. Comput Fluids 25(1):77–93
231. Casanellas L, Ortín J (2011) Laminar oscillatory flow of Maxwell and oldroyd-b fluids: theoretical analysis. J Non-Newton Fluid Mech 166(23–24):1315–1326
232. Casanellas L, Ortín J (2012) Experiments on the laminar oscillatory flow of wormlike micellar solutions. Rheol Acta 51(6):545–557
233. Mohyuddin MR, Götz T (2005) Resonance behaviour of viscoelastic fluids in Poiseuille flow in the presence of a transversal magnetic field. Int J Numer Meth Fluid 49(8):837–847
234. Torralba M, Castrejón-Pita JR, Castrejón-Pita AA, Huelsz G, del Río JA, Ortín J (2005) Measurements of the bulk and interfacial velocity profiles in oscillating Newtonian and Maxwellian fluids. Phys Rev E 72(1):016308
235. Torralba M, Castrejón-Pita AA, Hernández G, Huelsz G, del Río JA, Ortín J (2007) Instabilities in the oscillatory flow of a complex fluid. Phys Rev E 75(5):056307
236. Letelier M, Siginer DA, Caceres M (2002) Pulsating flow of viscoelastic fluids in tubes of arbitrary shape, Part I: longitudinal field. Int J Non-Lin Mech 37(2):369–393
237. Siginer DA, Letelier M (2002) Pulsating flow of viscoelastic fluids in tubes of arbitrary shape, Part II: secondary flows. Int J Non-Lin Mech 37(2):395–407
238. Beavers GS (1976) The free surface on a simple fluid between cylinders undergoing torsional oscillations, Part II: experiments. Arch Ration Mech Anal 62(4):323–352
239. Nikuradse J (1926) Untorsuchungen über die Geschwindigteitsverteilung in turbulenten Stromungen. Thesis, Göttingen, V.D.I.-Forsch, p. 281.
240. Prandtl L (1926) Über die ausgebildete turbulenz. Verfahren diese Zweite Internationale Kongress für Technische Mechanik, Zürich ["Turbulent flow," NACA Technical Memo 435, 62–75, 1927]
241. Prandtl L (1927) Über den Reibungswiderstand stromenderluft, Ergeb. Aerodyn. Versuch., Göttingen, III series
242. Einstein HA, Li H (1958) Secondary currents in straight channels. Trans Am Geophys Union 39:1085–1088
243. Brundrett E, Baines WD (1964) The production and diffusion of vorticity in duct flow. J Fluid Mech 19(3):375–394
244. Perkins HJ (1970) The formation of streamwise vorticity in turbulent flow. J Fluid Mech 44:721–740
245. Huser A (1992) Direct numerical simulation of turbulent flow in a square duct. PhD thesis, Department of Aerospace Engineering Sciences, University of Colorado
246. Hoagland LC (1960) Fully developed turbulent flow in straight rectangular ducts; secondary flow, its cause and effect on the primary flow. ScD thesis, Department of Mechanical Engineering, MIT
247. Leutheusser HJ (1963) Turbulent flow in rectangular ducts. ASCE J Hydraul Div 89:1–19
248. Gessner FB, Jones JB (1965) On some aspects of fully-developed turbulent flow in rectangular channels. J Fluid Mech 23:689–713
249. Launder BE, Ying WM (1972) Secondary flows in ducts of square cross-section. J Fluid Mech 54(2):289–295
250. Hinze JO (1973) Experimental investigation on secondary currents in the turbulent flow through a straight conduit. Appl Sci Res 28:453–465
251. Demuren AO, Rodi W (1984) Calculation of turbulence-driven secondary motion in non-circular ducts. J Fluid Mech 140:189–222
252. Nagata K, Hunt JCR, Sakai Y, Wong H (2011) Distorted turbulence and secondary flow near right-angled plates. J Fluid Mech 668:446–479

253. Bradshaw P (1987) Turbulent secondary flows. Annu Rev Fluid Mech 19:53
254. Mellor GL, Herring HJ (1973) A survey of the mean turbulent field closure models. AIAA J 11(5):590–599
255. Launder BE, Reece GJ, Rodi W (1975) Progress in the development of a Reynolds stress turbulence closure. J Fluid Mech 68:537–566
256. Hinze JO (1975) Turbulence. McGraw-Hill, New York, NY
257. Speziale CG (1987) On non-linear K-l and K-ε models of turbulence. J Fluid Mech 178:459–475
258. Yoshizawa A (1984) Statistical analysis of the deviation of the Reynolds stress from its eddy-viscosity representation. Phys Fluids 27(6):1377–1388
259. Yoshizawa A (1987) Statistical modeling of a transport equation for the kinetic energy dissipation rate. Phys Fluids 30(3):628–632
260. Shimomura Y, Yoshizawa A (1986) Statistical analysis of anisotropic turbulent viscosity in a rotating system. J Phys Soc Jpn 55(6):1904–1917
261. Nisizima S, Yoshizawa A (1987) Turbulent channel and couette flows using an anisotropic k-epsilon model. AIAA J 25(3):414–420
262. Yakhot V, Orszag SA (1986) Renormalization group analysis of turbulence. I. Basic theory. J Sci Comput 1(1):3–51
263. Rubinstein R, Barton JM (1990) Non-linear Reynolds stress models and the renormalization group. Phys Fluids 2(8):1472–1477
264. Speziale CG, So RMC, Younis BA (1992) On the prediction of turbulent secondary flows, NASA-ICASE Report No. 92-57
265. Lai YG, So RMC (1990) On near-wall turbulent flow modeling. J Fluid Mech 221:641–673
266. Speziale CG (1991) Analytical methods for the development of Reynolds stress closures in turbulence. Annu Rev Fluid Mech 23:107–157
267. Kolmogorov AN (1941) Local structure of turbulence in incompressible viscous fluid for very large Reynolds number. Dokl Akad Nauk SSSR 30:299–303
268. Barnes HA, Hutton JF, Walters K (1989) An introduction to rheology. Elsevier, Amsterdam
269. Einstein A (1906) Eine neue Bestimmung der Moleküledimensionen. Annal Phys 19:289–306
270. Einstein A (1911) Berichtigung zu meiner Arbeit: Eine neue Bestimmung der Moleküledimensionen. Annal Phys 34:591–592
271. Einstein A (1956) Investigations on the theory of the Brownian movement. Dover, New York, NY
272. Batchelor GK, Green JT (1972) The hydrodynamic interaction of two small freely moving spheres in a linear flow field. J Fluid Mech 56:375–400
273. Batchelor GK, Green JT (1972) The determination of the bulk stress in a suspension of spherical particles to order c^2. J Fluid Mech 56:401–427
274. Batchelor GK (1977) The effect of Brownian motion on the bulk stress in a suspension of spherical particles. J Fluid Mech 83:97–117
275. Krieger IM (1963) A dimensional approach to colloid rheology. Trans Soc Rheol 7:101–110
276. Krieger IM (1972) Rheology of monodisperse lattice. Adv Colloid Interface Sci 3:111–136
277. Stickell JJ, Powell RL (2005) Fluid mechanics and rheology of dense suspensions. Annu Rev Fluid Mech 37:129–149
278. Leighton D, Acrivos A (1987) The shear-induced migration of particles in concentrated suspensions. J Fluid Mech 181:415–439
279. Morris JF, Boulay F (1999) Curvilinear flows of non-colloidal suspensions: the role of normal stresses. J Rheol 43(5):1213–1237
280. Zarraga IE, Hill DA, Leighton DT (2000) The characterization of the total stress of concentrated suspensions of non-colloidal spheres in Newtonian fluids. J Rheol 44(2):185–220
281. Parsi F, Gadala-Maria F (1987) Fore-and-aft asymmetry in a concentrated suspension of solid spheres. J Rheol 31(8):725–732

282. Brady JF, Morris JF (1997) Microstructure of strongly-sheared suspensions and its impact on rheology and diffusion. J Fluid Mech 348:103–139
283. Wilson HJ (2005) An analytic form for the pair distribution function and rheology of a dilute suspension of rough spheres in plane strain flow. J Fluid Mech 534:97–114
284. Sierou A, Brady JF (2001) Accelerated Stokesian dynamics simulations. J Fluid Mech 448:115–146
285. Sierou A, Brady JF (2002) Rheology and microstructure in concentrated non-colloidal suspensions. J Rheol 46(5):1031–1056
286. Boyer F, Pouliquen O, Guazzelli É (2011) Dense suspensions in rotating-rod flows: normal stresses and particle migration. J Fluid Mech 686:5–25
287. Coutourier É, Boyer F, Pouliquen O, Guazzelli É (2011) Suspensions in a tilted trough: second normal stress difference. J Fluid Mech 686:26–39
288. Dbouk T, Lobry L, Lemaire E (2013) Normal stresses in concentrated non-Brownian suspensions. J Fluid Mech 715(1):239–272
289. Singh A, Nott PR (2003) Experimental measurements of the normal stresses in sheared Stokesian suspensions. J Fluid Mech 490:293–320
290. Joseph DD, Fosdick RL (1973) The free surface on a liquid between cylinders rotating at different speeds, Part I. Arch Ration Mech Anal 49(5):321–380
291. Joseph DD, Beavers GS, Fosdick RL (1973) The free surface on a liquid between cylinders rotating at different speeds, Part II. Arch Ration Mech Anal 49(5):381–401
292. Beavers GS, Joseph DD (1975) The rotating-rod viscometer. J Fluid Mech 69(3):475–511
293. Serrin J (1959) Mathematical principles of classical fluid mechanics (monograph). In: Truesdell C (ed) Handbuch der Physik, vol VIII/1. Springer, Berlin, pp 125–263
294. Joseph DD (1973) Domain perturbations: the higher order theory of infinitesimal water waves. Arch Ration Mech Anal 51(4):295–303
295. Siginer DA (1984) Free surface on a simple fluid between rotating eccentric cylinders, Part I: analytical solution. J Non-Newton Fluid Mech 15:93–109
296. Siginer DA, Beavers GS (1984) Free surface on a simple fluid between rotating eccentric cylinders, Part II: experiments. J Non-Newton Fluid Mech 15:109–122
297. Siginer DA (1984) General Weissenberg effect in free surface rheometry, Part I: analytical considerations. J Appl Math Phys 35(4):545–558
298. Siginer DA (1984) General Weissenberg effect in free surface rheometry, Part II: experiments. J Appl Math Phys 35(5):618–633
299. Wineman AS, Pipkin AC (1966) Slow viscoelastic flow in tilted troughs. Acta Mech 2 (1):104–115
300. Tanner RI (1970) Some methods for estimating the normal stress functions in viscometric flows. Trans Soc Rheol 14(4):483–508
301. Sturges L, Joseph DD (1975) Slow motion and viscometric motion, Part V: the free surface on a simple fluid flowing down a tilted trough. Arch Ration Mech Anal 59(4):359–387
302. Siginer DA (1991) Viscoelastic swirling flow with free surface in cylindrical chambers. Rheol Acta 30(2):159–175
303. Siginer DA, Knight RW (1993) Swirling free surface flow in cylindrical containers. J Eng Math 27:245–264
304. Deboeuf A, Gauthier G, Martin J, Yurkovetsky Y, Morris JF (2009) Particle pressure in a sheared suspension: a bridge from osmosis to granular dilatancy. Phys Rev Lett 102:108301
305. Prasad D, Kytomaa H (1995) Particle stress and viscous compaction during shear of dense suspensions. Int J Multiphas Flow 21(5):775
306. Phung TN, Brady JF, Bossis G (1996) Stokesian dynamics simulation of Brownian suspensions. J Fluid Mech 313:181–207
307. Yurkovetsky Y, I (1997) Statistical mechanics of bubbly liquids. II. Behavior of sheared suspensions of non-Brownian particles, PhD thesis, California Institute of Technology
308. Garland S, Gauthier G, Martin J, Morris J (2013) Normal stress measurements in sheared non-Brownian suspensions. J Rheol 57(1):71–89

309. Segré G, Silberberg A (1961) Radial particle displacements in Poiseuille flow of suspensions. Nature 189:209–210
310. Segré G, Silberberg A (1962) Behaviour of macroscopic rigid spheres in poiseuille flow Part 1. Determination of local concentration by statistical analysis of particle passages through crossed light beams. J Fluid Mech 14:115–135
311. Segré G, Silberberg A (1962) Behaviour of macroscopic rigid spheres in poiseuille flow Part 2. Experimental results and interpretation. J Fluid Mech 14:136–157
312. Gadala-Maria F, Acrivos A (1980) Shear induced structure in a concentrated suspension of solid spheres. J Rheol 24(6):799–815
313. Eckstein EC, Bailey DG, Shapiro AH (1977) Self-diffusion of particles in shear flow of a suspension. J Fluid Mech 79:191–208
314. Leighton D, Acrivos A (1987) Measurement of shear-induced self-diffusion in concentrated suspensions of spheres. J Fluid Mech 177:109–131
315. Nott PR, Brady JF (1994) Pressure-driven flow of suspensions: simulation and theory. J Fluid Mech 275:157–199
316. Brady JF, Bossis G (1985) The rheology of concentrated suspensions of spheres in simple shear flow by numerical simulation. J Fluid Mech 155:105–129
317. Brady JF, Bossis G (1988) Stokesian dynamics. Annu Rev Fluid Mech 20:111–157
318. Durlofsky LJ, Brady JF (1989) Dynamic simulation of bounded suspensions of hydrodynamically interacting particles. J Fluid Mech 200:39–67
319. Hampton RE, Mammoli AA, Graham AL, Tetlow N, Altobelli SA (1997) Migration of particles undergoing pressure-driven flow in a circular conduit. J Rheol 41(3):621–640
320. Phan-Thien N, Fang Z (1996) Entrance length and pulsatile flows of a model concentrated suspension. J Rheol 40(4):521–529
321. Karnis A, Goldsmith HL, Mason SG (1966) The kinetics of flowing dispersions: concentrated suspensions of rigid particles. J Colloid Interface Sci 22(6):531–553
322. Koh CJ (1991) Experimental and theoretical studies on two-phase flows. PhD thesis, California Institute of Technology
323. Koh CJ, Hookham P, Leal LG (1994) An experimental investigation of concentrated suspension flows in a rectangular channel. J Fluid Mech 266:1–32
324. Abbott JR, Tetlow N, Graham AL, Altobelli SA, Fukushima E, Mondy LA, Stephens ST (1991) Experimental observations of particle migration in concentrated suspensions: couette flow. J Rheol 35(5):773–795
325. Hookham P (1986) Concentration and velocity measurements in suspensions flowing through a rectangular channel. PhD thesis, California Institute of Technology
326. Sinton SW, Chow AW (1991) NMR flow imaging of fluids and solid suspensions in Poiseuille flow. J Rheol 35(5):735–773
327. Phillips RJ, Armstrong RC, Brown RA, Graham AL, Abbott JR (1992) A constitutive model for concentrated suspensions that accounts for shear-induced particle migration. Phys Fluid A 4:30–40
328. Chow AW, Sinton SW, Iwamiya JH, Stephens TS (1994) Shear-induced migration in couette and parallel-plate viscometers: NMR imaging and stress measurements. Phys Fluid A 6:2561–2676
329. Richardson JF, Zaki WN (1954) Sedimentation and fluidization: Part I. Trans Inst Chem Eng 32:35–47
330. MacDonald MJ, Muller SJ (1996) Experimental study of shear-induced migration of polymers in dilute solutions. J Rheol 40(2):259–283
331. Mills P, Snabre P (1995) Rheology and structure of concentrated suspensions of hard spheres. Shear induced particle migration. J Phys II 5:1597–1608
332. Subia SR, Ingber MS, Mondy LA, Altobelli SA, Graham AL (1998) Modelling of concentrated suspensions using a continuum constitutive equation. J Fluid Mech 373:193–219
333. Krishnan GP, Beimfohr S, Leighton DT (1996) Shear-induced radial segregation in bidisperse suspensions. J Fluid Mech 321:371–393

334. L'Huillier D (2009) Migration of rigid particles in non-brownian viscous suspensions. Phys Fluids 21:023302
335. Nott PR, Guazzelli E, Pouliquen O (2011) The suspension balance model revisited. Phys Fluids 23:043304
336. Blanc F, Lemaire E, Meunier A, Peters F (2013) Microstructure in sheared Non-Brownian concentrated suspensions. J Rheol 57(1):273–293
337. Brady JF (2001) Computer simulation of viscous suspensions. Chem Eng Sci 56 (9):2921–2926
338. Dratler DI, Schowalter WR (1996) Dynamic simulation of suspensions of non-Brownian hard spheres. J Fluid Mech 325:53–77
339. Drazer G, Koplik J, Khusid B, Acrivos A (2002) Deterministic and stochastic behaviour of non-Brownian spheres in sheared suspensions. J Fluid Mech 460:307–335
340. Chen S, Doolen GD (1998) Lattice Boltzmann method for fluid flows. Annu Rev Fluid Mech 30:329–364
341. Hill RJ, Koch DL, Ladd AJC (2001) The first effects of fluid inertia on flows in ordered and random arrays of spheres. J Fluid Mech 448:213–241
342. Glowinski R, Pan TW, Hesla TI, Joseph DD (1999) A distributed Lagrange multiplier fictitious domain method for particulate flows. Int J Multiphas Flow 25(5):755–794
343. Singh P, Hesla TI, Joseph DD (2003) Distributed Lagrange multiplier method for particulate flows with collisions. Int J Multiphas Flow 29(3):495–509
344. Nguyen N-Q, Ladd AJC (2002) Lubrication corrections for lattice-Boltzmann simulations of particle suspensions. Phys Rev E 66:046708
345. Kromkamp J, van den Ende D, Kandhai D, van der Sman R, Boom R (2006) Lattice Boltzmann simulation of 2D and 3D non-Brownian suspensions in couette flow. Chem Eng Sci 61(2):858–873
346. Maxey MR, Patel BK (2001) Localized force representations for particles sedimenting in stokes flow. Int J Multiphas Flow 27(9):1603–1626
347. Lomholt S, Maxey MR (2003) Force-coupling method for particulate two-phase flow: stokes flow. J Comput Phys 184(2):381–405
348. Dance SL, Maxey MR (2003) Particle density stratification in transient sedimentation. Phys Rev E 68:031403
349. Yeo K, Maxey MR (2010) Simulation of concentrated suspensions using the force-coupling method. J Comput Phys 229(6):2401–2421
350. Yeo K, Maxey MR (2010) Dynamics of concentrated suspensions of non-colloidal particles in couette flow. J Fluid Mech 649:205–231
351. Laun HM (1994) Normal stresses in extremely shear thickening polymer dispersions. J Non-Newton Fluid Mech 54:87–108
352. Tehrani M (1996) An experimental study of particle migration in pipe flow of viscoelastic fluids. J Rheol 40(6):1057–1077
353. Karnis A, Mason SG (1966) Particle motions in sheared suspensions. XIX. Viscoelastic media. Trans Soc Rheol 10:571–592
354. Gauthier F, Goldsmith HL, Mason SG (1971) Particle motions in non-Newtonian media II. Poiseuille flow. Trans Soc Rheol 15:297–330
355. Jefri MA, Zahed AH (1989) Elastic and viscous effects on particle migration in plane Poiseuille flow. J Rheol 33(5):691–708
356. Binous H, Phillips RJ (1999) Dynamic simulation of one and two particles sedimenting in a viscoelastic suspension of FENE dumbbells. J Non-Newton Fluid Mech 83:93–130
357. Lunsmann WJ, Genieser L, Armstrong RC, Brown RA (1993) Finite element analysis of steady viscoelastic flow around a sphere in a tube: calculations with constant-viscosity models. J Non-Newton Fluid Mech 48(1–2):63–99
358. Walters K, Tanner RI (1992) The motion of a sphere through an elastic liquid. In: Chhabra RP, Dekee D (eds) Transport processes in bubbles, drops and particles. Hemisphere, New York, NY

359. Solomon MJ, Muller SJ (1996) Flow past a sphere in polystyrene-based boger fluids: the effect on the drag coefficient of finite extensibility, solvent quality and polymer molecular weight. J Non-Newton Fluid Mech 62(1):81–94
360. Chilcott MD, Rallison JM (1988) Creeping flow of dilute polymer solutions past cylinders and spheres. J Non-Newton Fluid Mech 29:381–432
361. Harlen OJ (1990) High Deborah number flow of a dilute polymer solution past a sphere falling along the axis of a cylindrical tube. J Non-Newton Fluid Mech 37(2–3):157–173
362. Tiefenbruck G, Leal LG (1980) A note on rods falling near a vertical wall in a viscoelastic liquid. J Non-Newton Fluid Mech 6(3–4):201–218
363. Chiba K, Song K-W, Horikawa A (1986) Motion of a slender body in quiescent polymer solutions. Rheol Acta 25(4):380–388
364. Liu YJ, Joseph DD (1993) Sedimentation of particles in polymer solutions. J Fluid Mech 255:565–595
365. Kim S (1986) The motion of ellipsoids in a second-order fluid. J Non-Newton Fluid Mech 21 (2):255–269
366. Brunn P (1977) Interaction of spheres in a viscoelastic fluid. Rheol Acta 16(5):461–475
367. Riddle MJ, Narvaez C, Bird RB (1977) Interactions between two spheres falling along their line of centers in a viscoelastic fluid. J Non-Newton Fluid Mech 2(1):23–35
368. Joseph DD, Liu YJ, Poletto M, Feng J (1994) Aggregation and dispersion of spheres falling in viscoelastic liquids. J Non-Newton Fluid Mech 54:45
369. Gheissary G, van den Brule BHAA (1996) Unexpected phenomena observed in particle settling in non-Newtonian media. J Non-Newton Fluid Mech 67:1–18
370. Phillips RJ (1996) Dynamic simulation of hydrodynamically interacting spheres in a quiescent second-order fluid. J Fluid Mech 315:345–365
371. Binous H, Phillips RJ (1999) The effect of sphere-wall interactions on particle motion in a viscoelastic suspension of FENE dumbbells. J Nonnewton Fluid Mech 85:63–92
372. Blake JR (1971) A note on the image system for a stokeslet in a no-slip boundary. Proc Camb Phil Soc 70:303–310
373. Joseph DD, Liu YJ (1993) Orientation of long bodies falling in a viscoelastic liquid. J Rheol 37:961–983
374. Huang PY, Feng J, Hu HH, Joseph DD (1997) Direct simulation of the motion of solid particles in couette and Poiseuille flows of viscoelastic fluids. J Fluid Mech 343:73–94
375. Huang PY, Joseph DD (2000) Effects of shear thinning on migration of neutrally buoyant particles in pressure driven flow of Newtonian and viscoelastic fluids. J Non Newton Fluid Mech 90:159–185
376. Asmolov ES (1999) The inertial lift on a spherical particle in a plane Poiseuille flow at large channel Reynolds number. J Fluid Eng 381:63–87
377. Ramachandran A, Leighton DT (2008) The influence of secondary flows induced by normal stress differences on the shear-induced migration of particles in concentrated suspensions. J Fluid Mech 603:207–243
378. Zrehen A, Ramachandran A (2013) Demonstration of secondary currents in the pressure-driven flow of a concentrated suspension through a square conduit. Phys Rev Lett 110:018306
379. Belt RJ, Daalmans ACLM, Portela LM (2011) Experimental study of particle driven secondary flow in turbulent pipe flows. J Fluid Mech 709:1–36
380. Hetsroni G (1989) Particle-turbulence interaction. Int J Multiphas Flow 15(5):735–746
381. Gore RA, Crowe CT (1991) Modulation of turbulence by a dispersed phase. J Fluid Eng 113 (2):304–307
382. Huber N, Sommerfeld M (1994) Characterization of the cross-sectional particle concentration distribution in pneumatic conveying systems. Powder Tech 79:191–210
383. Lee SL, Durst F (1982) On the motion of particles in turbulent duct flows. Int J Multiphas Flow 8(2):125–146
384. Sommerfeld M (1990) Numerical simulation of the particle dispersion in turbulent flow: the importance of particle lift forces and particle/wall collision models. In: Celik I, Hughes D,

Crowe CT, Lankford D (eds) Numerical methods for multiphase flows, vol 91. ASME, New York, NY, pp 1–18
385. Saffmann PG (1965) The lift on a small sphere in a shear flow. J Fluid Mech 22:385–400
386. Mei R (1992) An approximate expression for the shear lift force on a spherical particle at finite Reynolds number. Int J Multiphas Flow 18(1):145–147
387. Matsumoto S, Saito S (1970) Monte Carlo simulation of horizontal pneumatic conveying based on the rough wall model. J Chem Eng Jpn 3:223–230
388. Sommerfeld M (1992) Modelling of particle-wall collisions in confined gas particle flows. Int J Multiphas Flow 18(6):905–926
389. Tsuji Y, Shen NY, Morikawa Y (1991) Lagrangian simulation of dilute gas-solid flows in a horizontal pipe. Adv Powder Tech 2:63–81
390. Oesterle B, Petitjean A (1993) Simulation of particle-to-particle interactions in gas-solid flows. Int J Multiphas Flow 19(1):199–211
391. Huber N, Sommerfeld M (1998) Modelling and numerical calculation of dilute-phase pneumatic conveying in pipe systems. Powder Tech 99:90–101
392. Launder BE, Spalding DB (1974) The numerical computation of turbulent flows. Comp Meth Appl Mech Eng 3:269–289
393. Li Y, McLaughlin JB, Kontomaris K, Portela L (2001) Numerical simulation of particle-laden turbulent channel flow. Phys Fluid 13(10):2957–2967
394. Flores AG, Crowe KE, Griffith P (1995) Gas-phase secondary flow in horizontal, stratified and annular two-phase flow. Int J Multiphas Flow 21(2):207–221
395. Jayanti S, Hewitt GF, White SP (1990) Time-dependent behaviour of the liquid film in horizontal annular flow. Int J Multiphas Flow 16(6):1097–1116
396. Jayanti S, Hewitt GF (1996) Response of turbulent flow to abrupt changes in surface roughness and its relevance in horizontal annular flow. Appl Math Model 20:244–251
397. Dykhno LA, Williams LR, Hanratty TJ (1994) Maps of mean gas velocity for stratified flows with and without atomization. Int J Multiphas Flow 20(4):691–702
398. Williams LR, Dykhno LA, Hanratty TJ (1996) Droplet flux contributions and entrainment in horizontal gas–liquid flows. Int J Multiphas Flow 22(1):1–18
399. Taitel Y, Dukler A (1976) A model for predicting flow regime transitions in horizontal and near horizontal gas-liquid flow. AIChE J 22:47–55
400. Van't Westende JMC, Belt RJ, Portela LM, Mudde RF, Oliemans RVA (2007) Effect of secondary flow on droplet distribution and deposition in horizontal annular pipe flow. Int J Multiphas Flow 33(1):67–85
401. Lin TF, Jones OC, Lahey RT, Block RT, Murase M (1985) Film thickness distribution for gas–liquid annular flow in a horizontal pipe. Physicochem Hydrodyn 6:179–195
402. Young J, Leeming A (1997) A theory of particle deposition in turbulent pipe flow. J Fluid Mech 340:129–159
403. Suzanne C (1985) Structure de l'Écoulement Stratifié de Gaz et de Liquide en Canal Rectangulaire, Thèse de Docteur ès Sciences, Institut National Polytechnique de Toulouse
404. Nordsveen M, Bertelsen AF (1997) Wave induced secondary motions in stratified duct flow. Int J Multiphas Flow 23(3):503–522
405. Andrews DG, McIntyre ME (1978) An exact theory of non-linear waves on a Lagrangian-mean flow. J Fluid Mech 89:609–646
406. Langmuir I (1938) Surface motion of water induced by wind. Science 87:119–123
407. Nordsveen M, Bertelsen AF (1996) Waves and secondary flows in stratified gas/liquid duct flow. In: Grue J, Gjevik B, Weber JE (eds) Waves and non-linear processes in hydrodynamics. Kluwer Academic Publishers, Dordrecht
408. Siginer DA (2012) Isothermal tube flow of non-linear viscoelastic fluids, Part I: constitutive instabilities and the longitudinal field. Int J Eng Sci 56(7):111–126
409. Siginer DA (2011) Isothermal tube flow of non-linear viscoelastic fluids, Part II: transversal field. Int J Eng Sci 49(6):443–465

410. Rotta JC (1951) Statistische Theorie nichthomogener Turbulenz. Zeitschrift für Physik A Hadrons and Nuclei 129(6):547–572
411. Rotta JC (1951) Statistische Theorie nichthomogener Turbulenz. Zeitschrift für Physik A Hadrons and Nuclei 131(1):51–77
412. Rotta JC (1972) Recent attempts to develop a generally applicable calculation method for turbulent shear flow layers, AGARD-CP-93, North Atlantic Treaty Organization
413. Rotta JC (1972) Turbulente Strömungen: Eine Einführung in die Theorie und ihre Anwendung. B. G. Teubner Verlag, Stuttgart
414. Daily BJ, Harlow FH (1970) Transport equations in turbulence. Phys Fluids 13:2634–2649
415. Hanjalic K, Launder BE (1972) A Reynolds-stress model of turbulence and its application to thin shear flows. J Fluid Mech 52:609–638
416. Shir CC (1973) A preliminary numerical study of atmospheric turbulent flows in the idealized planetary boundary layer. J Atmos Sci 30:1327–1339
417. Cormack DE, Leal LG, Steinfeld JH (1978) An evaluation of mean Reynolds stress turbulence models: the triple velocity correlation. J Fluid Eng 100:47–54
418. Hanjalic K, Launder BE (1976) Contribution towards a Reynolds-stress closure for low-Reynolds-number turbulence. J Fluid Mech 74:593–610
419. Launder BE, Reynolds WC (1983) Asymptotic near-wall stress dissipation rates in a turbulent flow. Phys Fluids 26:1157–1158